THE RAILROAD

Recent Titles in
Greenwood Technographies

Sound Recording: The Life Story of a Technology
David L. Morton Jr.

Rockets and Missiles: The Life Story of a Technology
A. Bowdoin Van Riper

Firearms: The Life Story of a Technology
Roger Pauly

Cars and Culture: The Life Story of a Technology
Rudi Volti

Electronics: The Life Story of a Technology
David L. Morton Jr. and Joseph Gabriel

Computers: The Life Story of a Technology
Eric G. Swedin and David L. Ferro

THE RAILROAD

THE LIFE STORY OF A TECHNOLOGY

H. Roger Grant

GREENWOOD TECHNOGRAPHIES

GREENWOOD PRESS
Westport, Connecticut • London

Library of Congress Cataloging-in-Publication Data

Grant, H. Roger, 1943–
The railroad : the life story of a technology / H. Roger Grant.
p. cm. — (Greenwood technographies, ISSN 1549–7321)
Includes bibliographical references and index.
ISBN 0–313–33079–4 (alk. paper)
1. Railroads—United States—History. I. Title. II. Series.
TF23.G675 2005
385'.0973—dc22 2004028177

British Library Cataloguing in Publication Data is available.

Library of Congress Catalog Card Number: 2004028177
ISBN 0–313–33079–4
ISSN: 1549–7321

First published in 2005

Greenwood Press, 88 Post Road West, Westport, CT 06881
An imprint of Greenwood Publishing Group, Inc.
www.greenwood.com

Printed in the United States of America

The paper used in this book complies with the Permanent Paper Standard issued by the National Information Standards Organization (Z39.48–1984).

10 9 8 7 6 5 4 3 2 1

For S. M. Grant (1863–1955)
My Scottish grandfather whose father died in an
accident on the Highland Railway

Contents

Series Foreword

In today's world, technology plays an integral role in the daily life of people of all ages. It affects where we live, how we work, how we interact with each other, and what we aspire to accomplish. To help students and the general public better understand how technology and society interact, Greenwood has developed *Greenwood Technographies*, a series of short, accessible books that trace the histories of these technologies while documenting *how* these technologies have become so vital to our lives.

Each volume of the *Greenwood Technographies* series tells the biography or "life story" of a particularly important technology. Each life story traces the technology from its "ancestors" (or antecedent technologies), through its early years (either its invention or development) and rise to prominence, to its final decline, obsolescence, or ubiquity. Just as a good biography combines an analysis of an individual's personal life with a description of the subject's impact on the broader world, each volume in the *Greenwood Technographies* series combines a discussion of technical developments with a description of the technology's effect on the broader fabric of society and culture—and vice versa. The technologies covered in the series run the gamut from those that have been around for centuries—firearms and the printed book, for example—to recent inventions that have rapidly taken over the modern world, such as electronics and the computer.

While the emphasis is on a factual discussion of the development of the technology, these books are also fun to read. The history of technology is full of fascinating tales that both entertain and illuminate. The authors—all experts in their fields—make the life story of technology come alive, while also providing readers with a profound understanding of the relationship of science, technology, and society.

Introduction

The Railroad: The Life Story of a Technology covers one of the great inventions in modern history. It would be rails, initially wooden, then iron and later steel, that would bind together farm and factory and shatter the isolation of huge land masses. Some economic historians, in fact, suggest that during the nineteenth century the iron horse largely facilitated the industrial "takeoff" in North America, making possible an heavily urbanized society. It is hardly surprising that a significant (and extensively recorded) event in the history of the United States occurred on May 10, 1869, at a remote location in Utah Territory known as Promontory Summit, where an enthusiastic crowd gathered to watch the "wedding of the rails." The ceremonial spikes and hammers, wired to telegraph lines, made electrical connections that sent signals through the nation's telegraphic network. The ceremony proclaimed completion of the transcontinental railroad whose tracks wound through the rugged Sierra Nevada mountains, thrust across the central plains, and splayed across the East. The future of a wondrous age seemed at hand; the iron horse and the telegraph proclaimed progress to the citizenry. Undeniably these two remarkable technologies greatly contributed to economic, political, social, and cultural change both in North America and in other parts of the world.

If ever there has been organic growth with a technology, it has been with the railroad. This transport form has undergone *gestation*, *youth*, *maturity*, and *old age*. Yet, *old age* is somewhat misleading, since the drive for technological improvements has never ceased. Still, in more recent years outside forces have handicapped American railroad companies from exploiting betterments, largely because of inadequate financial resources caused by antiquated governmental and labor restrictions, ever-growing intermodal competition, and a degree of internal corporate ossification. But during the latter two decades of the twentieth century railroads underwent a rebirth of sorts that was brought about by fewer external controls, impressive technological breakthroughs, and more "think-outside-the-box" managerial attitudes.

The railroad is many things. It is track, motive power, rolling stock, signals, structures, and much more. And every aspect of railroad technology has experienced change. Mostly, these improvements and refinements have worked, making for a faster, safer, and more dependable form of intercity transportation. Practical individuals, both domestically and abroad and from both within and outside the industry, have promulgated advancements. Hardly surprisingly, as with most areas of technology, some schemes really came to naught, whether actually attempted or only contemplated. Examples abound. The narrow-gauge building craze that swept the United States following the Civil War failed to produce a standard alternative width between the rails. Similarly, nearly a century later plans to construct atomic-powered locomotives never got off the drawing boards.

Even though railroads are presently perceived by many to be less important because of the revolution in airplanes, automobiles, and trucks, freight railroads still remain essential for a postindustrial world economy, hauling a nearly infinite variety of raw materials and semi- or fully manufactured goods. And railroads continue to move people about. Although Amtrak in the United States and VIA in Canada are not large passenger carriers, heavily patronized, high-speed trains have thoroughly established themselves in parts of Europe and Asia. Yet even Amtrak, with its recently publicized *Acela Express* and *Metroliner* trains, which operate along Amtrak's active Northeast Corridor between Boston, Massachusetts, and Washington, D.C., regularly travel at speeds exceeding 100 miles per hour. There is a growing belief that in time much faster trains, whether conventional or highly radical, will link major population centers in the United States and in other parts of the world. Gridlock on the highways and in the skyways will be a driving force contributing to the best railroad transport options that are humanly possible. Even though during the *old age* phase a few pundits predicted the death of the railroad, any obituary is today completely out of date. All railroads will not become "rails-to-trails" recreational paths, items for museums, or mostly forgotten memories. Technology has changed that possibility, and technology will continue to keep railroads alive and serving the freight and passenger needs of millions.

The Railroad: The Life Story of a Technology is not designed to be an exhaustive account of the remarkable experiences with the iron horse. Rather, this work offers an overview of the development of railroads, largely in the context of the United States, suggesting wide-ranging changes that have happened and continue to occur. Hopefully, the complexities of railroad technology, past and present, can be better understood and appreciated.

Timeline

Mid-1700s	Primitive mine "waggonways" open in England and Germany.
1769	Joseph Kugnot operates his "steam carriage" in Paris, France.
1804	A steam-powered amphibious dredge makes its debut in Philadelphia, Pennsylvania. The Welsh Merthyr Tydvil Railroad hauls its first coal train.
1805	Beacon Hill Railroad, Boston, Massachusetts, opens.
1825	Stockton & Darlington Railway begins commercial operations in England.
1826	Granite Railway, Charleston, Massachusetts, handles its first stone cargoes.
1827	Construction of the Baltimore & Ohio Railroad begins.
1829	The steam locomotive *Stourbridge Lion* begins trials on the Delaware & Hudson Railroad.
1830	*Best Friend of Charleston* steam locomotive pulls its first train. *Tom Thumb*, an experimental steam locomotive, competes against horse-drawn train on Baltimore & Ohio Railroad.
1831	First "T" rail is laid on the Camden & Amboy Railroad in New Jersey. Matthias W. Baldwin launches the Baldwin Engine and Locomotive Company (later Baldwin Locomotive Works).

Late 1830s	Eight-wheel cars appear.
1840s	Robert Davidson creates a battery-powered locomotive in Scotland.
1847	New York State outlaws strap rails.
1851	The 450-mile New-York & Erie Railroad opens between Piermont and Dunkirk, New York, creating the world's longest railroad line. Electric telegraph is first used for train dispatching. First refrigerator car is introduced.
Mid-1850s	First steel rail is rolled.
1856	Theodore Woodruff receives a U.S. patent for a sleeping car with upper and lower berths.
1862	First U.S. mail car.
1864	George Pullman introduces the sleeping car *Pioneer*.
1868	James Eads uses first pneumatic caisson for bridge building.
1869	The first transcontinental railroad is completed. George Westinghouse tests his "straight air brake."
Early 1870s	First all-metal freight cars enter service on Prussian State Railways.
1871	Robert Fairle pronounces merits of narrow-gauge railroads.
1877	Billerica & Bedford Railroad becomes first 2-foot-gauge U.S. common carrier.
1878	Baker steam-heating apparatus for passengers cars introduced.
1879	First all-steel railroad bridge is completed at Glasgow, Missouri. Telephone first used for train dispatching.
1881	Installation of Edison incandescent lamps in passenger cars of the London, Brighton & South Coast Railway of England.
1883	Establishment of Railway Standard Time.
1885	Leo Daft operates experimental electric locomotive in New York City.
1887	Frank Sprague electrifies street railways in Richmond, Virginia. Interstate Commerce Commission (ICC) established.
1893	Safety Appliance Act enacted. *Theory and Construction of a Rational Heat Engine* by Rudolf Diesel is published.
1904	First all-steel passenger cars introduced on the Interborough Rapid Transit System in New York City. William McKeen devises a self-propelled gasoline passenger car.

1920s	Introduction of train air conditioning.
1925	First diesel-electric locomotive begins revenue service on the Jersey Central Railroad.
1927	First important installation of centralized traffic control (CTC).
1934	Inaugural run of the diesel-powered *Zephyr.*
Mid-1930s	Regular commercial trailer-on-flatcar (TOFC) service begins.
1938	Test trials of General Motors freight diesel freight locomotives (F units).
1939	Continuous welded rail (CWR) introduced.
1949	Rail diesel car (RDC) appears.
1950s	Introduction of "Big John" covered hopper cars. First remote-controlled locomotives.
1951	Erie Railroad begins its "four-way train-radio-telephone" network.
1954	Union Pacific Railroad installs first computer.
1956	Malcolm McLean introduces water freight container service.
1960s	Modern wave of railroad corporate mergers.
1971	Amtrak begins.
Late 1970s	RoadRailer service starts.
1980s	Freight containerization (container-on-flatcar, or COFC) becomes commercially popular.
1980	Staggers Act provides partial deregulation of U.S. railroads.
1981	*TGV* makes debut in France.
1984	Double-stack COFC service begins.
1988	CSX Transportation opens its computerized-based Dufford Control Center.
1990s	Adoption of AC traction for diesel-electric locomotives. Increased usage of remote-controlled locomotives.
2000s	Meglev development.

1

Gestation, 1800–1860

FIRST MOVEMENTS

For nearly 200 years the railroad in America has steadily evolved. Yet this magnificent means of long-distance transportation, both for freight and passengers, did not spring magically from the creative minds of native inventors as did the incandescent light bulb, the television set, and the personal computer. Although Americans contributed enormously to perfecting the railroad, British mechanics and tinkerers largely made the earliest developments. A free-flowing transfer of technology between Great Britain and the United States resulted in the formulation of arguably the most significant invention in the rise of the modern industrial state. Citizens soon altered their views about time, distance, and convenience. And the railroad seemed wonderfully suited for what that perceptive French visitor of the 1830s, Alexis de Tocqueville, aptly called the "restless temper" of Americans.

It is virtually impossible to date the "first" railroad in the world. But likely it occurred early in the mid-sixteenth century with the introduction of privately owned and operated tramways or "waggonways" that served the transport needs of English collieries. Unquestionably these tiny, primitive affairs fit the standard definition of a railroad: an overland right-of-way with a fixed path consisting of paired rails that are elevated to support self-guided vehicles rolling on flanged wheels (wheels with a projecting rim or collar). These new transport devices convincingly demonstrated that a flanged wheel running on a flatly laid rail encounters less friction than any other type of wheeled conveyance. For more than 200 years an assortment of Lilliputian carriers used draft animals, gravity, or both, and occasionally wind or water flow to propel coal-laden cars to their destinations, usually a nearby tidewater, river, or canal port. There these bulky minerals were reloaded onto aquatic transport. Since rails and other materials were nearly always made of wood, such pikes were commonly called "wooden railroads."

Just as it is difficult to date initial tramway operations in England, it is also hard to pinpoint when the first railroad appeared on American soil. A strong candidate would be the Beacon Hill Railroad (BHR) in Boston, Massachusetts, one of approximately twenty-five wooden railroads that before 1830 dotted the eastern seaboard. In 1805 this pygmy pike, which consisted of about one-fifth of a mile of track, began to shunt cars of dirt and rock so that a real estate development company could pare down historic Beacon Hill. At the base the firm sought to create land suitable for building commercial structures. In reality, then, the BHR was a temporary contractor's railway, designed to facilitate heavy "cut-and-fill" work. Indeed, for some time English builders had installed such railways during the construction of bridges, lighthouses, and tunnels. Although details about the short-lived BHR are few, at least one brief commentary exists: "There was a Rail Road running in a southwesterly direction from the top of the hill: It struck Cedar Street a little to the South of Mt. Vernon Street, and struck Charles Street on the east side . . . : It was used with a large pulley at the top fastened to each set of cars, and one set of cars went up while the others went down, both being attached together: There were branch rails at the top and the bottom: It would be difficult . . . to say how many men and teams there were" (Gamst 1992, 93).

Much more is known, however, about another Massachusetts wooden tramway, the Granite Railway (GR), sometimes called the Quincy Rail Road. In 1823 a patriotic organization, formed to honor the fiftieth anniversary of the famed Battle of Bunker Hill in the Revolutionary War, decided to erect a granite obelisk on Breed's Hill in Charleston, near Boston. In order to bring construction materials to the monument site, the practical response would be to load stone at a quarry in West Quincy and transport it overland to the Neponset River for placement on barges to Charlestown harbor. In early 1826 the General Court of Massachusetts awarded a charter to the GR for conducting "general transportation," although blocks of granite would be the *only* commodity hauled. By October this 3-mile tramway was in operation, functioning as its sponsors had expected, and it would exist for the next quarter century. A combination of horses and gravity provided locomotion, and the company's wooden cars on their iron wheels rumbled over wooden rails with iron caps (or "strap rails") that had been fastened to a "continuous foundation of granite blocks . . . [which rested] on a base of dry stonework" (Gamst 1997, 295). The GR possessed a lightly graded right-of-way, consisting of several sections with different inclinations to accommodate the hilly terrain. A variety of wooden viaducts carried the tracks over deep ravines.

Early on, commentators praised operations. In January 1829, William Jackson, a member of the Massachusetts Charitable Mechanics Association, told colleagues that "the construction, uses and advantages of the Quincy Rail Road, I presume, are familiar to most of those who hear me." Specifically, he noted that "the cars upon this road carry from four to five tons. Two cars are considered as a load for one horse, traveling at the rate of 3 miles an hour. This is for continued operations through the day. If the horse, however, is put to its utmost capacity, such as is frequently the case with the truck horses of this city [Boston], he can carry double this load." Concluded Jackson, "Many apprehensions were entertained that our severe frosts and deep snows would injure and obstruct its operations.—But the frosts of two winters have done no harm to the road; nor has the snow. . . . The success of his Rail Road is fortunate, not only because it accomplishes the object for which it was designed, but because it serves to give confidence to calculations for other and more extensive roads" (Jackson 1829, 12).

The success of the GR continued to be widely recognized. "This unique project first demonstrated in America the engineering advantages of rail transport," observed the American Society of Civil Engineers later in the nineteenth century, "and introduced many technical features, such as switches, the turntable and double-truck railway cars" (Vose 1884, 67–68). And, of course, both the long-term experiences of the British with tramways and the triumph of the Beacon Hill Railroad played an enormous role in the accomplishments of the GR.

The "tramway era," though, had largely run its course by the time the Granite Railway hauled its first car of heavy stone. Soon the prototypes for the "modern" railroad in the United States emerged, most notably the Baltimore & Ohio Railroad (B&O), Boston & Lowell Railroad, Boston & Worcester Rail Road, Charleston & Hamburgh Rail-Road (née South-Carolina Canal and Rail-Road), and Petersburg Railroad (of Virginia). Yet these pioneer carriers incorporated features of the vanishing tramways, including the basic track structure.

During what may be thought of as the "Demonstration Period," those decades immediately prior to the Civil War, fledgling railroads developed what became the accepted practices of line location and roadbed and rail construction. James Gadsden, a consultant to a group of promoters who in the 1830s wished to construct a steam-powered road between Charleston, South Carolina, and the Ohio River, made it clear in his report of 1836 that "*the most perfect Road* therefore, in its construction, must be that on a *horizontal plane*, and on a *straight line*, or on the shortest distance practicable between the two points to be connected" (Gadsden 1836, 7). This became the widely accepted axiom, which new building techniques made increasingly possible. When it came time to install the trackwork, the vast majority of the pioneer railroads opted to place it directly on graded surfaces. Gangs of workers would attack the landscape with picks, shovels, wheelbarrows, and carts to accomplished their assigned tasks. Often they would hitch husky teams of oxen to iron-tipped plows in order to loosen the earth. At times it would be necessary to dig deep cuts and occasionally to bore tunnels with the aid of black powder, always with the intended purpose of maintaining a gradient that was as level as possible. These rights-of-way mostly followed the contours of the ground, thus keeping the grading work to a minimum.

Yet not all early railroad builders believed that carving a right-of-way out of the terrain was the best approach. Several projects, including the ambitious New-York & Erie Railroad (NY&E), involved the installation of extensive timber pilings for the track foundation. "The adoption of piles in the construction of nearly two hundred miles of the road-way," explained NY&E President Eleazar Lord in the *Second Report of the Directors*, published in 1841, "is . . . deemed a great advantage over the ordinary method of grading." And he offered these pertinent details: "The piles used are generally of white oak, about twelve inches in diameter. They are driven by steam power, five feet apart from center to center, to such depth in all cases as to secure them from the effects of frost, and with such force as to leave them in no danger of settling under the pressure of any load. The tops of the piles being protected by timbers wide enough to cover them, it is supposed that they may endure about twelve or fifteen years." Concluded Lord, "They form a road of uniformly even surface throughout the year, which is not liable to be obstructed by snow, and while they continue sound, will be subject to little or no expense for repairs. The difference between a piled and graded road for annual repairs will, it is believed, be sufficient in five or six years to defray the entire cost of renewing the piles; while the expense of working such a road, owing to the constant evenness and good condition of the rails, will be very considerably less than is commonly incurred on graded roads" (New-York & Erie Railroad Company 1841, 15–16).

Why Eleazar Lord and his associates on the New-York & Erie opted for a piled roadway is not fully clear. Of course, by the 1840s standards for the creation of rights-of-way had not been wholly accepted. The earlier Charleston & Hamburgh had experienced a modicum of success with its pilings, although this strategy became mandatory over a section of swampy terrain in the South Carolina Low Country. Perhaps, too, NY&E leaders were influenced by their desire to attract more public enthusiasm for their sputtering project; they had committed themselves to building a nearly 450-mile railroad between the Hudson River, near New York City, and Dunkirk, New York, a port community on Lake Erie, and they needed the financial backing of area interests. Indeed, local suppliers of timber and their families, friends, and neighbors might become more supportive of this ambitious undertaking; after all, some residents of the Southern Tier region viewed the railroad as of benefit principally to the citizens of New York City and its environs. And there was another consideration. Since the railroad would traverse a territory that repeatedly experienced heavy rain and snow, these "stilts" would make the track structure "free from dangers of a graded road in consequence of the washing of the banks by flood and rains, and settling when set up in soft bottom," and "not liable to be obstructed by snow" (Grant 1994, 3).

Whatever the reasoning, the piling strategy failed miserably. Labor costs escalated, appropriate timbers were not always available, and the process of rot in the untreated wood set in quickly. In all, the NY&E spent about $1 million on this ill-fated endeavor. When the line finally opened between "the Ocean and the Lakes" in 1851, it possessed a traditional roadbed. This early passion for wood, however, did not prevent the company from the construction of massive stone bridges and viaducts, and, remarkably, most of them continue to be used today.

Although during the Demonstration Period railroad contractors found large timbers to be enormously practical, especially for bridge and trestle construction, builders swiftly decided that small pieces of wood made for excellent crossties or simply "ties." (The British terms of "sills" or "sleepers" were sometimes used.) Even primitive tramways usually had their wooden rails spiked to round crossties supported by some sort of dirt or rock ballast. Yet a number of pioneer roads, including one of the first common carriers, the B&O, preferred granite blocks to support their rails. At times this material was readily available and promoters, resembling those in Britain, thought about creating track structures that would last for "eternity." In the late 1820s and early 1830s the B&O tried two different support methods: the second proved to be more unsatisfactory than the first. Initially track gangs placed a wood stringer with its iron strap rail on top of regularly spaced cubical stone blocks. The chief engineer, however, then adopted an "improved" track design that consisted of longitudinal granite stringers laid end-to-end in a trough, with the strap iron rail fastened directly to the stone. One track between Relay (near Baltimore) and Ellicott City, Maryland, was laid with stone blocks, which measured about 12 by 18 inches, were a foot thick, and weighed about 200 pounds each; the second track to Ellicott City and both tracks from Ellicott City to Sykesville, Maryland, about 20 miles altogether, were laboriously and expensively built with the longitudinal *stone* stringers. At that point the B&O realized that a solid stone trackbed was the wrong approach. Besides being costly and slow to install, it turned out to be completely unsuitable for railroad service, particularly once steam locomotives came into usage, because stone made the track much too rigid. Officials did not reveal an undue pessimism with the stone sleeper failure; they took the matter in stride. As a B&O director later said, "Everything [then] was new, crude, and doubtful" (Dilts 1993, 122).

A consistently more successful method of railroad building involved the rails themselves. Experience gained from tramways revealed that it was practical to move rolling stock over wooden rails. Yet for durability it was quickly discovered that these members needed to be covered with iron straps nailed on them, hence the common term strap rails. But having so much wood meant almost immediate problems. Primitive, albeit wholly inadequate timber preservation techniques existed, usually involving the charring of the wood or using a tar and sand coating. While highly rot-resistant woods existed, principally cedar and cypress, the convenience of pine and certain hardwoods made them cheaper and more practical for most railroad builders. Installing the entire wood track members usually directly on top of earth hardly enhanced longevity. Yet, techniques emerged for assembling these track components that retarded decay. A civil engineer suggested that the wood used for rail stringers be "quartered or split through the heart," rather than square-hewed. Timbers fashioned this way had "less exposure to decay . . . contracts less by seasoning," and were less likely to "check," a natural process that opened the wood and permitted water to collect internally (Bianculli 2003, 88). Some builders opted for a coating of heated tar and oil.

As the 1830s began, American railroads, once more using British expertise and products, found "English rails" more suitable for their needs. Typical of rails installed were those received by the B&O in mid-1829; specifically, they were 0.5 inches thick, 2.25 inches wide, and 15 feet long. The iron was slightly rounded on top with holes punched every 18 inches and ends cut to fit together in a special joint. The most troubling dimensions involved the expense and availability of these essential building components.

Then, as railroads launched their regular freight and passenger operations, managers discovered the inherent weakness of strap rails. Increased train weights and speeds doomed strap-rail technology, for at times they caused strap rails to break apart. Even light trains, running at modest speeds, created a rolling action that occasioned spikes to loosen and rail to curl. The detached rails, dubbed "snake heads," with their sharp pieces of metal, could spear through the floors of the wooden cars, halting and even derailing the train much as snags and sawyers might impale and sink a river boat. Personal injuries, and even deaths, commonly resulted. Not surprisingly, strap rails became the subject of legislative debates, promoting New York lawmakers in 1847 to outlaw their usage on common carriers; they demanded employment of the best technology. Other states followed suit. Still, financially weak companies, mostly on the frontier, used these readily available and cheap secondhand strap rails for a decade or two longer. In Wisconsin and in the Upper Peninsula of Michigan, for example, they could occasionally be found as late as the 1870s. A distinguished Russian civil engineer who visited the United States during the strap-rail era offered his assessment: "About eight years of experience in the existence of such roads in the United States have shown that this system is not beneficial, and that it may be permitted only in extreme instances where, due to a total lack of capital, [metal] bar rails cannot be had, or where the road is intended only for slow transport of light freight without the use of locomotives" (Melnikov n.d.).

In the 1830s and 1840s much better rail became available. *All-metal* wrought iron rails came into wider usage, and they set the standard until the advent of durable steel rail after the Civil War. While weight and length lacked standardization, these rails customarily weighed 36 to 50 pounds to the linear yard and came in pieces that measured 12, 15, or 18 feet. Moreover, specific configurations of the earliest solid iron rails also varied, including "H," "U," and "T" shapes. It would be the latter, however, that rapidly gained dominance. Even though in the late eighteenth century Englishman William Jessop likely devised the first T-rail, it was Robert Stevens, builder of the Camden & Amboy Rail Road and Transportation Company and a creative American civil engineer, who played a fundamental role in development of T-rail. He reputedly whittled the shape of his version from a block of wood while he sailed from the United States to England on a mission to buy iron rail for his New Jersey road. The Stevens design allowed greater contact with the wheel and better traction for the locomotive than Jessop's pattern. Furthermore, this improved model could be set directly on the crosstie and fastened down with a hook-headed iron spike, of which Stevens also fashioned a prototype. Once in Great Britain with his model T-rail he commissioned a leading manufacturer to handle his experimental order, and then in 1831 this rail was laid on the Camden & Amboy. But not all of the railroad sported this unique rail. This resulted not because of faulty design but rather due to the high cost of production and delivery, factors that slowed its widespread adoption among American railroads.

During the Demonstration Period and even later, railroad builders debated what should be the appropriate distance between the iron-capped or all-metal rails. In fact, this issue was not fully resolved in the United States until long after the "Railway Age" had fully been established. For the most part, early railroad builders accepted the British notion that the optimal railway gauge should be 4 feet, 8½ inches. After all, tramways and the first major commercial carrier, the Liverpool & Manchester, commonly featured either 4 feet, 8 inches or 4 feet, 8½ inches. This gauge had long been customary for British carts and wagons and could be traced back to Roman times.

The only major exception to this "standard" gauge in the British Isles occurred on the Great Western Railway. Its iconoclastic builder, Isambard Kingdom Brunel, considered 4 feet, 8½ inches grossly inadequate for the speeds that he envisioned, expecting that trains on his road between London and southwestern England would operate at 50 to 60 miles per hour rather than the 25 to 30 miles per hour common to other British pikes. When he planned the Great Western in the early 1830s, he specified the width to be a staggering *7 feet*. This turned out to be the widest gauge ever used in a railway designed for general service.

The British commitment to the 4-foot, 8½-inch gauge had a profound effect on gestating American railroads. Many of these carriers selected "English gauge" because they owned English-made locomotives. Nevertheless, there were important variations. The Camden & Amboy selected a gauge of 4 feet, 10 inches, while the Delaware & Hudson initially used an odd 4-foot, 3-inch gauge. Yet, most other railroads in Delaware, Maryland, New York, and Pennsylvania featured standard gauge. In Ohio, however, railroads were built usually to 4 feet, 10 inches, probably because the first important carrier in the state, the Mad River & Lake Erie Railroad, acquired locomotives equipped with that gauge from a New Jersey manufacturer. A few roads in the Buckeye state even had a 5-foot, 4-inch distance between their rails. In the South the popular gauge became 5 feet, except in North Carolina and Virginia where railroads preferred standard gauge. Selection by the pioneer South-Carolina Rail Road and Canal Company of the 5-foot gauge likely influenced other carriers in Dixie.

Just as Isambard Kingdom Brunel opted for an unusually broad gauge, there would be American examples. But the gauge was not a whopping 7 feet; instead, it measured 6 feet. The instigator of this width would be the New-York & Erie Railroad. As with its failed piled roadway experiment, this company once again made an expensive error in construction, taking decades to rectify it. Officials believed that there existed several advantages for using what would soon be called "Erie Gauge." Since their rails and ties would not rest on stilts, they worried about how trains would manage some of the road's stiff grades. These men concluded that the only practical way to overcome these obstacles would be to operate locomotives of "enormous weight," perhaps 20 tons, a mass thought so great that only a track of broad gauge could support them. Some executives, too, opined that in time the NY&E might need to operate their freight and passenger trains in "squads" (or sections), and that their numbers could be kept fewer if the heaviest possible locomotive were employed. The savings, therefore, with these more efficient operations would surely be considerable. Yet the *real* explanation for a gauge of 6 feet had nothing to do with technology. There was the quietly expressed concern that freight traffic might be diverted by other railroads to Baltimore, Philadelphia, or other destinations if the NY&E shared the same gauge. Subsequently, the NY&E, reorganized in the 1860s as the Erie Railway, was able through affiliates and connections to be a key link in a network of broad-gauge lines that by 1870 stretched from New York to St. Louis.

North America had become a land of track-gauge variations. By the eve of the Civil War the breakdown of American railway gauges was as follows: the 4-foot, 8½-inch gauge accounted for 53.3 percent of the track; 5-foot gauge for 21.8 percent; 4-foot, 10-inch gauge for 9.9 percent; 5-foot, 6-inch gauge ("Canadian gauge") for 8.7 percent; and 6-foot gauge for the remaining 5.3 percent. The remaining 1 percent consisted of modest variations to these five widths.

Although the battle of the gauges was far from resolved by the time of the Civil War, an important legislative event did much to give the nation a standard rail width. On July 1, 1862, Congress approved "an Act to aid in the construction of a railroad and telegraph line from the Missouri River to the Pacific coast, and to secure to the Government the use of the same for postal, military, and other purposes." An important provision of this Pacific Railway Act, which underwrote construction of the Union Pacific and Central Pacific railroads, involved gauge: "The track upon the entire line of railroad and branches shall be of uniform width, to be determined by the President of the United States, so that, when completed, cars can be run from the Missouri River to the Pacific Coast." President Abraham Lincoln, however, preferred a 5-foot gauge, but various railroad leaders and politicians wanted 4 feet, 8½ inches, and within a year Congress repealed the section that gave the president authority over selecting gauge. The replacement provision required "that the gauge of the Pacific railroad and its branches throughout their whole extent, from Pacific coast to the Missouri River shall be, and hereby is, established at four feet eight and one-half inches" (Taylor and Neu 1956, 55).

STEAM APPEARS

What would propel these infant railways? By the late 1820s that question had become the principal topic of discussion and even debate among railroad enthusiasts. For centuries operators of wooden tramways relied for locomotion on either gravity or animal power, usually horses. Yet the latter had limitations. For one thing, capital expenses were considerable. In the Horse Age, these animals commanded hefty prices, often exceeding thousands of dollars. Moreover, horses required considerable amounts of food, stabling, and overall care. Horses also were susceptible to injury and disease, including the highly contagious epizootic lymphangitis, a respiratory illness. And they needed to be rested periodically. Speeds, too, were limited. Although this was usually not a major concern for bulk freight movements, the advent of regularly scheduled passenger service meant for relatively slow intercity transport times.

During these formative years, inventors toyed with alternatives to animal power. A few explored the possibilities of wind. In some ways this was not surprising; after all, wind power, centerpiece of the first industrial revolution, appeared long before the stationary steam engine. It would be in Holland, world center of wind technology, that early in the seventeenth century a wind-powered wagon (*zeilwagen* or "sail wagon") designed by the inventor Simon Stevin, which reportedly held several passengers, moved as fast as 20 miles per hour along the beach at low tide between Scheveningen and Noordwijk. Arguably this experimental contraption could claim to be the first land vehicle to move under power other than that supplied by animals or humans. But difficulties with managing sails, frequent absences of favorable breezes, or the appearance of erratic gales and the prevailing winds from the southwest (the return trip to Schevenigen from Noordwijk had to be made with animal power) caused this source of locomotion, when attempted on the hard-packed beach or on primitive rails, to be rapidly discarded. Still, it can be argued that Stevin's creation with its wide wheels, primitive steering apparatus, and sail functioned surprisingly well.

By the eighteenth century, steam power seemed promising for propelling vehicles. As early as 1769, Frenchman Nicholas Joseph Cugnot fashioned a steam carriage that he operated in Paris. Thirty-five years later the American tinkerer Oliver Evans created his strange and imaginative *Oruktor Amphibolos* (amphibious digger) in Philadelphia. This 21-ton steam-powered machine, aptly described as "a barge like, steam-powered, amphibious dredge," snorted and clanked through city streets and made several trips on the Delaware River. Although his creation was hardly a monumental success, Evans fervently believed that "carriages propelled by steam will be *in general use*, as well for the transportation of passengers as goods, traveling at the rate of 15 miles an hour, or 300 miles per day" (Dilts 1993, 82–83).

The true prototype of the steam-power railway locomotive, however, appeared in Wales and not Pennsylvania. It was Richard Trevithick, a Cornish mining engineer, who built presumably the world's first steam locomotive. He assembled a power device that sported two axles, a single cylinder, and a return-flue, high-pressure boiler. In February 1804 the Trevithick contraption pulled several tiny freight cars of the industrial Merthyr Tydvil Railroad. The experimental train consisted of a 10-ton load of coal and moved over the iron-capped rails at an average speed of 5 miles per hour.

Much more experimentation occurred, mostly taking place in the British Isles. Although some technological disappointments occurred, progress was rather steady and impressive. By the early 1820s several steam locomotives had been successfully operated. Surely the earliest master of steam locomotive construction was George Stephenson, who along with his gifted son, Robert, built in the mid-1820s *Locomotion No. 1* for the Stockton & Darlington Railway, the world's first modern railroad. On its maiden run, *Locomotion No. 1* pulled a "substantial" train of cars at speeds of about 20 miles per hour, or four times that of Trevithick's train of twenty-one years earlier. In 1829 Stephenson constructed an even more powerful steam locomotive, which he appropriately named the *Rocket*. That year the Stephenson machine performed well at the famous Rainhill Trials on the Liverpool & Manchester Railway, demonstrating during a week of grueling tests its superiority over several competing steam locomotive types. Moreover, Stephenson had introduced three distinguishing design features: a multitube, horizontal boiler; a forced draft that enhanced steaming speed and volume; and pistons directly connected to the four coupled driving wheels.

Events in Great Britain inspired Americans. A leading illustration was the work of Colonel John Stevens, a college-trained lawyer and a talented inventor and mechanic. Widely read and fully cognizant of events abroad, he had long been convinced of the usefulness of steam energy and the practicability of the railroad. Earlier Stevens had gained acclaim for his steam-powered vessel that had triumphantly navigated the Hudson River. Next he built a demonstration railroad at his sprawling estate in Hoboken, New Jersey. Moving along this half-mile of circular track was a steam locomotive of his own design that featured a small, upright boiler. Thus Stevens built and operated the first steam locomotive on American rails. Although not producing immediate, commercial results, his creative labors, widely discussed in engineering and transportation circles, gave impetus to the steam-powered railway movement. Moreover, residents of the New World could now ride behind a steam-powered vehicle that offered them a far more tangible demonstration of steam's potential power than elusive, contemporary newspaper accounts.

The next landmark in bringing mechanical propulsion to the gestating American railroad involved the first steam locomotive to operate in commercial service. Rather than employing equipment designed and built by Stevens or some other domestic genius, Britain became the source of "the first locomotive that ever turned a driving wheel on a railroad track in America." John B. Jervis, chief engineer of the Delaware and Hudson Company (D&H), a firm that profitably operated both an important coal-carrying canal and a 16-mile, largely gravity-powered railroad for the movement of coal from mines near Honesdale, Pennsylvania, to the waterway, in late 1827 dispatched his assistant Horatio Allen to England to purchase several steam locomotives. After arriving in "the land of railroads and in the atmosphere of coal smoke," Allen arranged for the purchase of three locomotives from a leading manufacturer, Foster, Rastrick and Company. During 1829 these iron horses arrived in the United States, including what became the renowned *Stourbridge Lion*. But once trials began on the D&H, it became apparent that this pricey ($3,000) locomotive was too heavy for the track structure. Reported Jervis, "We have had the Lion on the road from the Basin to about half across [and] at this point one of the capts [iron straps] began to fail and we run [*sic*] her back a little." The engine "stood on the road all right," but "it will be necessary to give additional support to the bridge." Unfortunately, this "magnificent machine," which featured a horizonal multitubular boiler (boiler with multiple steam pipes, or flues), a tall smokestack, a separate firebox, and four large driving wheels, was impractical for its assignment. Soon management relegated the *Stourbridge Lion* to the less dramatic role of stationary steam engine on one of the firm's included planes. Horses continued to power rail operations (Larkin 1990, 28–31).

Much happier experiences with steam locomotives soon took place on other early railroads. American manufacturers rapidly appeared and their products, often with designs heavily influenced by British models, began to ply domestic rails. On Christmas Day 1830 the *Best Friend of Charleston* took the honor of being the first commercial locomotive designed and constructed in the United States. A product of the West Point Foundry, located in New York City, this 4.5-ton steamer, which cost an impressive $4,000, began its brief career as the principal motive power for the South Carolina Rail Road & Canal Company. Although the *Best Friend of Charleston* performed adequately, it shockingly blew up in June 1831 when a fireman, who was killed by the explosion, had foolishly held down the safety value because the sound of its hissing steam annoyed him.

The chief company engineer of the South Carolina road, Horatio Allen, Jervis's former colleague, held high hopes for this replacement technology. "There is no reason to expect any material improvements in the breed of horses in the future," he concluded, "while in my judgment, the man is not living who knows what the breed of locomotives is to place at command" (*New York Times* August 9, 1877). Allen's judgement was solid. Nearly a century later, a South Carolina newspaper editor argued that "the locomotive of 100 years ago will stand out in marked contrast to the enormous passenger locomotives of today that are capable of drawing heavy trains at a speed of more than 60 miles per hour." He added, "It is well to hark back a hundred years occasionally and gather fresh inspiration through a demonstration of the rapid strides that have been made in all lines of human endeavor" (*Edgefield* [SC] *Advertiser* November 14, 1928).

Although there was much public hoopla associated with the first running of the *Best Friend of Charleston*, surely one of the most talked-about races in nineteenth-century American history generated even greater excitement. The place was Maryland, and the railroad was the newly opened initial segment of the Baltimore & Ohio. In 1830 Peter Cooper, a gluemaker from New York City and an inveterate inventor, created a small, experimental 1-ton locomotive that he appropriately dubbed the *Tom Thumb*. Although details of its mechanical makeup are obscure, it is known that a steam-powered piston propelled a chain mechanism that alternately bypassed and engaged cogged wheels and raked them forward. The process created continuous forward motion. Soon, though, Cooper used a more conventional arrangement, namely, he connected rod, crank, and spur and pinion gears. The result was a more dependable machine that could travel much faster. In August the *Tom Thumb* showed off its stuff on the outbound trip from Baltimore when it covered 13 miles at speeds of up to 18 miles per hour. And it was on the return trip that "the race" allegedly occurred. Years later a contemporary observer remembered:

> But the triumph of this Tom Thumb engine was not altogether without a drawback. The great stage proprietors of the day were Stockton and Stokes; and on this occasion a gallant gray of great beauty and power was driven by them from town, attached to another car on the second track . . . and met the engine at the Relay House on its way back. From this point it was determined to have a race home; and, the start being even, away went horse and engine, the snort of the one and the puff of the other keeping time and tune. At first the gray had the best of it, for *his* steam would be applied to the greatest advantage on the instant, while the engine had to wait until the rotation of the wheels set the blower to work. The horse was perhaps a quarter of a mile ahead when the safety value of the engine lifted and the thin blue vapor issuing from it showed an excess of steam. The blower whistled, the steam blew off in vapory clouds, the pace increased, the passengers shouted, the engine gained on the horse, soon it lapped him—the silk was plied—the race was neck and neck, nose and nose—then the engine passed the horse, and a great hurrah hailed the victory. But it was not repeated; for just at this time, when the gray's master was about giving up, the band which drove the pulley, which drove the blower, slipped from the drum, the safety value creased to scream, and the engine for want of breath began to wheeze and pant. In vain Mr. Cooper, who was his own engineman and fireman, lacerated his hands attempting to replace the band upon the wheel; in vain he tried to urge the fire with light wood; the horse gained on the machine, and passed it; and although the band was presently replaced, and steam again did its best, the horse was too far ahead to be overtaken. (Dilts 1993, 94)

Momentarily the leadership of the Baltimore & Ohio had serious reservations about the utility of steam locomotion. Yet the following year the railroad conducted extensive trials with much larger power. The 3.5-ton *York*, which had been constructed in the Pennsylvania town of the same name, turned out to be the star performer. An upright boiler and 30-inch drivers characterized this locomotive. The conclusion of the B&O was that the *York* was fully practical; it proved to be more dependable and managed to chug along at a blistering 30 miles per hour.

In the early 1830s other largely experimental steam locomotives made their debut. In the summer of 1831 the *DeWitt Clinton*, an impressive machine built by the West Point Foundry with a horizontal boiler rather than a vertical one the company had fashioned for the *Best Friend of Charleston*, plied the iron-clad rails of the 17-mile Mohawk & Hudson Railroad between the New York cities of Albany and Schenectady. Its four driving wheels measured 54 inches in diameter, and it sported a "tender" of novel design. The lower part of this trailing unit was actually a container for the water supply, arguably the first "water-bottom tank." The *DeWitt*

Throughout its corporate life the Baltimore & Ohio Railroad took pride in having hosted in 1830 Peter Cooper's experimental 1-ton *Tom Thumb*. For its historic trip near Baltimore, the locomotive steamed along effectively before mechanical difficulties caused it to lose an impromptu race with a horse-powered train. The upright boiler developed 75 horsepower at a speed of 10 miles per hour and could pull a small car at 12 miles per hour. Courtesy of the author.

Clinton revealed signs of engineering developments that would have lasting impact.

Another steam locomotive of significance was the *John Bull*, built in 1831 for the Camden & Amboy. Although constructed in Great Britain by the Stephensons, this piece of motive power would later have one of the first enclosed cabs, surely a delight to engineer and fireman alike during inclement weather. Moreover, Matthias W. Baldwin, a cunning craftsman who had inspected the newly arrived *John Bull*, later incorporated some of its features into the products of his recently launched locomotive-manufacturing business in Philadelphia, destined to become the industry giant. By the eve of the Civil War, a thousand engines had left the Baldwin shops and thousands more were to be built.

Even with the rapid and impressive advancement in steam locomotive technology, some railroad leaders remained unconvinced that horse-powered trains had become wholly obsolete. Just as individuals at the B&O had initial reservations, so did organizers of the New-York & Erie. "Animal power may be considered the natural power of the country," observed a company spokesman in 1833, "and on long routes, where great inequalities in the amount of travel and transport will occur; where the commodities to be conveyed, instead of presenting a regular supply, will probably amount to many times as much some months as others, the use of horses may be expected, for a time at least, to be practically cheaper than steam." The company also envisioned feeder lines, owned by the public or private investors, to employ horses. "Branch rail-ways . . . may be constructed at a moderate expense, by counties, towns, villages, associations, and even by individuals, where those adapted to the use of steam could not be undertaken" (Wright 1833, 7). These points of view held logic, but only for a brief time. Advances in steam propulsion would shortly be even more breathtaking and overcome the objections.

But technological advances took place in an unique *American* environment. What became apparent, even as early as the trials of the *Stourbridge Lion* on the Delaware & Hudson, was that the developing railroad lines in the United States, with few exceptions, would be much more lightly built than those in Great Britain and later in most of Europe. After all, English promoters had access to large pools of capital, enjoyed more highly developed metalworking technologies, and possessed considerable numbers of trained engineers. Repeatedly they strove to create the best railroads that were humanly possible. Earthen fills, stone bridges, and brick-lined tunnels characterized British railroad construction and resulted in mostly level rights of way. These expenditures made considerable sense: the kingdom was small, densely populated, and fully involved in the "second" industrial revolution of steam-powered industries. On the other hand, American railroads, which sprawled out over thousands of square miles, commonly possessed steep grades, excessive curvature, and flimsy wooden bridges. Tracks were almost universally cheaply built, and they were thus more dangerous.

To meet these challenges in the United States, the original 0-4-0 wheel arrangement was speedily abandoned. (Steam locomotives are commonly referred to by their classification in the Whyte system, which uses the number of pilot-truck wheels, drivers, and trailing-truck wheels to identify most kinds of steam motive power. Thus the "American Standard" or "8-wheeler" is a 4-4-0, which means it has a four-wheel pilot truck, two pairs of coupled drivers, and no trailing trucks. See the Appendix for more information on the Whyte System.) Rather a new type of wheel arrangement was introduced. Beginning in 1832 six-wheel truck locomotives or 4-2-0s entered service and gained the appellation of "One Armed Billies" for their single set of drivers and their maker, William Norris. These engines benefited considerably from their swiveling leading wheels or trucks, greatly improving their ability to navigate the all-too-common undulating and roughly built American tracks. Then, in about 1839, the Utica and Schenectady Rail Road started to rebuild its original 4-2-0 locomotives as eight-wheel or 4-4-0 engines. Indeed, this became a popular mechanical improvement: at modest cost, a second set of driving wheels could be placed in front of the firebox, a betterment that more than doubled tractive effort. Specialists considered the 4-4-0 to have approximately 60 percent more pulling strength than the 4-2-0. By the 1840s the 4-4-0 had become the locomotive of choice for domestic railroads; it proved to be agile and powerful in operation. The enormous success of the 4-4-0 design allowed early American builders like Baldwin, Norris, and Thomas Rogers to export them back to England and to Europe. Railroad technology transfer had now become truly two-directional.

Not surprisingly, not all American locomotives by the 1840s featured either the 4-2-0 or 4-4-0 wheel arrangement. As coal traffic increased on several railroads, motive power needed to be stronger. In the early part of the decade, locomotive manufacturer Ross Winans, who possessed an amazing talent for locomotive and equipment design, developed an 0-8-0 type engine for the Baltimore & Ohio. These locomotives, which weighed about 23 tons, featured what had become the standard horizontal boiler powered by

Early in the twentieth century a postcard maker selected an image of a re-creation of the first train that chugged over the rails of the Mohawk & Hudson Railroad. Unlike the *Tom Thumb*, this engine revealed the direction the modern locomotive would take: horizontal boiler, steam dome, and fuel-carrying tender. Courtesy of the author.

eight rather small drivers. Known as "Mud Diggers" because of their propensity to derail on sharp curves and for their side rods that almost touched the ground as they revolved, these Winan engines managed to pull heavy coal and merchandise trains. Yet, as one official complained, they had "become too unwieldy and too liable to get out of repair, to be efficient" (Dilts 1993, 365). Still, the 0-8-0 wheel arrangement found contemporary supporters.

At the end of the 1840s, the always-creative Winans produced a better 0-8-0. His first "Camel" or "Camelback" locomotives contained a large cab for the engineer located on top of the boiler and a huge firebox cantilevered from its rear. The latter feature gave the fireman direct access to the coal supply carried in the tender. These slow-moving steamers weighed more than 25 tons and like the Mud Diggers had a 0-8-0 wheel arrangement, yet they were more powerful, rugged, and reliable. But unlike Winan's earlier creations and those locomotives of other designs used elsewhere, these Camelbacks were the first *coal*-burning engines produced in quantity for an American railroad. They also possessed a distinctly different appearance: "The most peculiar engine in use in the United States," concluded a contemporary authority. "In every detail of construction," it was "alike peculiar and in the strongest possible contrast with the proportions, arrangement, and workmanship of the standard American engine" (Clark and Colburn 1861, 50). Ungainly though they were, Camelback-style locomotives worked well, not only on the B&O but also on other largely coal-hauling railroads.

The process of technology transfer between the cradle of railroading and the New World did not mean that everything was adopted or modified. The premier example was the "atmospheric railways" or "ropes of air railways," an early, albeit short-lived attempt to replace steam locomotives with stationary steam power. In the 1840s three British railways tested the possibilities of *compressed* air: the Dublin & Kingstown Railway, London & Croydon Railway, and South Devon Railway. The ingenious concept involved propelling trains in this fashion: a piston ran in a tube between the rails with the direct train connection being made by a mechanism that included a greased leather flap that sealed a slot along the top of the pipe. Stationary pumping engines exhausted the tube ahead of the piston, producing a vacuum. Air then admitted through the leather valve behind the piston created the "push" or motive power. Although the technology worked, it might best be described as a brilliant near-miss, and these roads abandoned this alternative after a year or so. The popular story (a true "urban legend") that rats ate the leather valves is a myth, even though the English public commonly believed that these pesky rodents ruined an exciting new railroad technology.

Although technological dead-ends were part of the early development in railroading on both sides of the Atlantic Ocean, the quest for increased speed of locomotives remained consistent. Better track structures—most notably solid iron rails spiked to hardwood ties—allowed faster movement of trains. Everyone seemingly wanted "to travel like the wind," and American locomotive builders responded to this challenge.

In the late 1840s a notable illustration of the demand for speed occurred. As the Vermont Central Railroad approached completion, its president sent a representative to the Baldwin Locomotive Company in Philadelphia with a request. The railroad wanted a locomotive that could handle a passenger train at a speed of 60 miles per hour, and offered $10,000 for such a mile-a-minute piece of motive power. Baldwin was game. Design and construction began quickly, and in 1849 company craftsmen completed the *Governor Paine*.

The completed locomotive was novel. It had one pair of driving wheels, seventy-eight inches in diameter, placed back of the firebox and another pair, albeit smaller and unconnected, located directly in front of the firebox. The cylinders were approximately 17 inches in diameter and had 20-inch stroke, being installed horizontally between the frames and the boiler. The

By the mid-nineteenth century the steam locomotive had acquired refinements that would characterize it for decades, including a headlight, bell, and enclosed cab. Yet this circa 1860 view of Portland & Kennebec Railway's American-type (4-4-0) locomotive reveals technological features that became obsolete, most notably being a wood burner with a balloon smoke stack and having running gears that are inside connected. Courtesy of the author.

reason for placing the cylinders at about the middle of the boiler was to reduce the lateral motion of the locomotive. The bearings on the two rear axles were so designed that by means of a lever, a part of the weight of the engine that was usually carried on the wheels in front of the firebox could be transferred to the driving axle, thus enhancing tractive effort. And a four-wheel truck assembly carried the front of the engine. The *Governor Paine* was fast. In one test run, the locomotive sprinted along a mile of track in an impressive 43 seconds. (Slightly modified versions of these engines went to the Pennsylvania Railroad, where they reached a speed of 4 miles in 3 minutes.) Yet a major problem developed: the locomotive lacked sufficient adhesion to the rails. This flaw prompted the Vermont Central to rebuild the *Governor Paine* with two sets of standard driving wheels. And, of course, faster, more powerful locomotives were in the future, produced not only by Baldwin but also by other manufacturers.

CARS AND CARRIAGES

Early on, railroad builders turned considerable attention to the development of rolling stock. After all, locomotives of whatever design were created to pull a variety of revenue-producing equipment. In the tradition of the slow-moving animal- and gravity-powered tramways, small, albeit practical freight cars appeared on the first common carriers. Initially, operators employed the four-wheel British style, but speedily abandoned it (except for coal "jimmies") for the more stable double-truck car. And unlike contemporary British railways, American carriers did not prefer to haul merchandise in canvas-covered "open" cars. Weather extremes, especially in the North, and often long distances between major population centers promoted the rapid development of permanently covered cars. Moreover, the overall size of American freight equipment was bigger, at times considerably larger than that found on English and European roads. This "jumbo-sized" rolling stock, commonly with a 10-ton capacity by the 1860s, became an American hallmark that foreign visitors repeatedly noticed.

During the Demonstration Period, casual trackside observers might have noted the presence of freight rolling stock, perhaps sensing that cars were getting larger, but they surely did not realize that a variety of technological improvements made for easier, heavier, and faster movements. As early as 1826 Ross Winans, the later locomotive-design genius, demonstrated with a test car the advantages of the friction-wheel bearings over conventional plain bearings that allowed for the moving of heavy payloads. Rapid wear and the tendency of wheel sets to wander back and forth made this particular Winans triumph much less so, but in the 1830s significantly improved wheel assemblies appeared, commonly featuring leaf springs, journal boxes (for lubrication), and better bearings. The tendency increasingly became to replace hardwood with iron in freight rolling stock from trucks to frames. Still, until about 1870, carmakers continued to use considerable quantities of timber products, largely because of their cheapness, abundance, and workability. During these decades, after a few years of experience, building components that were originally believed to be necessary proved to be superfluous. Indeed, simplification is a common feature of developing technologies.

Betterments also involved something as mundane as how rolling stock would be physically linked. The first couplings were merely an iron bar with a hole in each end or an iron chain fastened to a pin on the end of each car. Chain couplings, however, made starts and stops so jerky that early trainmen often put chunks of wood between the cars to hold them apart and to keep the chains taut. But as cars became heavier, the link-and-pin coupler, known to later railroaders as the "Lincoln" pin, came into general prominence. This replacement device consisted of a slotted wrought-iron drawbar fastened to the end of the car, into which a long oval link was inserted and held in place by an iron pin that a brakeman dropped into a hole in the drawbar once the coupling had occurred.

This form of standardization aided the interchange of equipment between railroads. Initially not much thought was given to the matter of coupling. Since companies lacked connecting partners, having a uniform way to connect cars was not a concern. But as the rail network increased, a freight car or even a passenger coach might travel from road to road.

Although the coal car, flatcar, and box or "merchanise" car became popular products of rolling-stock builders, the Demonstration Period also saw the appearance of speciality equipment designed to meet the needs of particular shippers. In the 1830s containerization, commonly associated with railroad rolling stock of the last quarter of the twentieth century, made a debut. Several Eastern roads, including the Camden & Amboy, used baggage container cars. Large, wooden boxes were placed on conventional flatcars, allowing for the protection of trunks from the elements and from easy theft. In a similar fashion, in 1849 the Richmond, Fredericksburg & Potomac Railroad began placing on flatcars a dozen crates each that held small quantities of freight, what later became known as less-than-carload (LCL) shipments. And on the Pennsylvania State Works, an ingenious publicly financed network of railroad, incline plane and canal that stretched between Philadelphia and Pittsburgh, small canal boats, which were loaded onto flatcars for their rail journeys, entered regular, albeit temporary revenue service.

Since transporting foodstuffs was a vital part of freight traffic in the preindustrialized United States, it is understandable that railroads gave special attention to the shipment of livestock. Although cattle drives had existed since colonial times—drovers, for example, had for decades herded steers from rural Delaware northward to Wilmington and Philadelphia—rail transport offered the faster, safer, and healthier delivery of table-destined animals. What appeared by the 1850s were "compromise cars," boxcars that featured open-barred side doors and end ventilator doors to be used for livestock and alternative solid doors that safely enclosed grain or merchandise. Following the Civil War, the familiar side-slatted roofed cattle car became common when cattle shipments to national meat-packing centers substantially increased.

Constant improvements, design variations, and changing needs in freight rolling stock really did not create major problems for the gestating railroad industry. When cars were built during this wooden period, their life expectancy was never anticipated to be particularly great. Although the time length in service varied, it was common to "retire" a piece of equipment after ten years or so. On English and European railroads, however, rolling stock (or "burden wagons"), which was usually better built, was expected to remain in service for a much longer time.

From virtually their inception, railroads hauled both "hogs and humans." And some of the early passengers were not certain if they traveled any more comfortably than did barnyard animals. Just as the first freight cars rolled on four wheels, the earliest American passenger cars were simple four-wheel carriages. It is probably not true that there was a *direct* technology transfer between stagecoach bodies and railway coaches, namely, it is unlikely that workers actually mounted bodies of former stagecoaches on flanged wheels. Yet the general appearance of the passenger cars of the 1830s correctly suggests that established carriage makers introduced the first railway passenger equipment. But soon some manufacturers produced what were known as four-wheel "compartment" cars, merely uniting several units into a single vehicle, and their creations suggested the appearance of what became the standard passenger coach of the pre–Civil War era. By the late 1830s the "modern" eight-wheel car, longer and wider than previous passenger equipment and frequently with an open platform at each end, made its debut on the Boston & Providence Railroad and was subsequently widely employed.

The popular reaction to these cars, irrespective of design, was that the travel experience was often unpleasant. These pioneer pieces of passenger equipment were spartan, offering for example little in the way of seating amenities: "The seat ran along each side, like those of the omnibus, and the coaches equally destitute of any and every other appliance for the comfort or conveyance of the traveler, other than to sit down and 'hang-on'—if he could," observed a rider about 1840 on the Northern Cross Rail Road, the first common carrier in Illinois. "A sudden lurch of the coach would often slide a sitter half the length of the coach and land him, or her, with a gruesome bump in the middle of the floor" (Grant 2004, 8). The hard, low-back seats, which were so common, surely reminded passengers of church rather than home.

But quality improved, at times noticeably. In the mid-1840s an observer noted that the car builders had made major upgrades, at least for the passenger trains that operated on New York's Auburn & Rochester Railroad: "The cars are each 28 feet long and 8 feet wide. The seats are well stuffed and admirably arranged—with arms for each chair, and changeable backs that will allow the passenger to change 'front to rear' by a maneuver unknown in military tactics." And he added, "The size of the cars forms a pleasant room, handsomely painted, with floor matting, with windows secured from jarring, and with curtains to shield from the blazing sun." Moreover, the ride itself was better. "These cars are so hung on springs, and are of such large size, that they are freed from most of the jar, and especially from the swinging motion so disagreeable to most railroads" (quoted in White 1978, 69).

As the century progressed, the plain lines, square windows, and common lack of architectural ornamentation on passenger cars yielded to more rounded appearances, large-arched windows, and clerestory roofs (for improved ventilation). Additional decorative features also graced each coach. And interiors, too, improved markedly. Not only did seats become more comfortable, but heating also improved. During the winter months, wood-burning iron stoves were installed. And over the years it became common for shop employees to remove several seats in the middle of the car for the stove placement in order to equalize the heating effects. No longer was the stove (if it existed at all) found at the end of the car. After 1860 coal-fired car stoves gradually replaced wood-burning units. Illumination got better, too, but it was a gradual process. By the 1850s candles gave way to oil lamps. But before the age of petroleum (kerosene) there was no cheap lamp oil, explaining the longevity of inexpensive candles. Indeed, this traditional form of lighting was simple, smokeless, and relatively safe. In the 1830s and for some time afterwards, toilets were not always included, yet in the mid-1830s they became available on a few cars. When added to a coach, they were primitive, nothing more than a wooden box with a circular hole cut in the top for a seat, paralleling the ubiquitous home privy. Human waste dropped directly on the tracks, much to the annoyance of maintenance-of-way workers and anyone else who walked these metropolitan corridors.

Whether they built freight cars or passenger coaches, craftsmen dominated the manufacturing process. There was no high-volume production with workmen performing only one specialized step at a time, characteristic of an assembly-line organization of later years. The antebellum car workers operated in small teams, understanding and executing all of the steps required to assemble a particular piece of rolling stock. These artisans started on the frame and continued to add parts until the roof was in place; surely they took considerable pride in their final accomplishments. Undoubtedly, too, they realized that their handiwork helped to set the nation in motion and knit the country more tightly together.

STATIONS

The earliest railroad leaders in the United States surely experienced some satisfaction in their construction of trackside support structures, most of all freight and passenger depots. Yet these buildings may not have been constructed at the time service began. Apparently pioneer railroaders did not fret much about depot design or construction. They needed to build their lines, provide suitable locomotives and rolling stock, and recruit reliable employees. The rail owners' goal, of course, was to begin operations quickly in order to generate much-needed revenues. Usually lacking extra funds because traffic was light, these individuals decided to use available buildings, if possible, for depot-related services.

During its formative years, the activities of the Baltimore & Ohio indicate the typical thinking of officials about structures. Not long after the spindly little road reached Ellicott City, Maryland, in May 1830, the company decided that its passengers could use the recently finished Patapsco Hotel near its end-of-track. After all, riders could fend for themselves. In fact, this had been the long-established custom for stagecoach, canal, and lake and river boat passengers. Operators of stage lines, for example, did not usually own their station facilities; rather, private proprietors of hotels, stores, and taverns served patrons. If necessary, travelers made their own arrangements for food and lodging. Yet the infant B&O still felt the need to erect a building in Ellicott City, and carpenters constructed a type of multipurpose depot (and in the mid-1850s this structure began to serve passengers as well). The company wished to use its limited resources for a facility that could accommodate its local workforce and, most of all, its freight shipments. The explanation is simple: merchandise—valuable LCL freight—required protection from weather and thieves.

When railroads ordered depots to be erected, they might become attractive affairs. While "country" stations surely did not tax the building know-how of carpenters—some sported basic wood framing and featured simple board-and-batten exteriors—urban terminals were noticeably different. Usually these structures were either the "temple" or "train-barn" type. The former were modeled after small classical Greek temples. The tracks ran through the side of these usually frame structures under one galley of the peristyle, and waiting rooms and offices occupied the remainder of the enclosed space. The latter allowed trains to run *through* the center of the depot, and the architecture was virtually indistinguishable from that of a large country barn. Both structures nearly always contained roof ventilators, permitting the wood smoke of locomotives to escape.

Yet prototypes of the great monuments to the transportation boom appeared prior to the Civil War. In 1847 the Old Colony Railroad, for example, erected on Kneeland Street in Boston, Massachusetts, an impressive largely brick depot complete with an adjoining trainshed. Designed by a local architect, this three-story structure was surely one of the most fully equipped railroad facilities of the day. It provided a smoking room, barbershop, newsstand, and other amenities for passengers. Surely by mid-century travelers began to expect both a depot and a facility that adequately served their needs. Urban residents, too, started to realize that the depot was the unofficial gateway to their communities, and they wanted a building that reflected the growing importance of their hometowns. Visitors often judged a place by the quality of its depot, and they would continue to do so as the railroad network moved beyond its gestation stage.

The appearance and development of railroads electrified Americans. By the mid-nineteenth century the public commonly believed that the tyranny of isolation and distance that separated them would be overcome through a series of technologies that involved the iron horse. The ever-improving steam locomotive made the greatest impact on actual travel and the national psyche. "The locomotive engine has in twenty years," wrote railroad expert and trade publisher Henry Varnum Poor in 1851, "become the great agent of civilization and progress, the most powerful instrument for good the world has yet reached, and become the most effective messenger for proclaiming peace on earth and good will to men." And he opined, "The age of locomotion is the era of progress—Wherever the railway extends, knowledge and civilization advance in geometrical ratio" (quoted in Ward 1986, 29).

Although the "civilizing" impact of railroads really knew no geographical bounds, the developing railroads in the United States were noticeably *different* from those in Great Britain and on the European continent. American builders thought and acted in terms of the *immediate* and the *practical*. "The key to the evolution of the American railway is the contempt for authority displayed by our engineers, and the untrammeled way in which they invent and applied whatever they thought would answer the best purpose, regardless of precedent," concluded a writer for the *Magazine of American History* in June 1891. "When we began to build our railways in 1831, we followed English patterns for a short time. Our engineers soon saw that unless vital changes were made, our money would not hold out, and our railway system would be very short" (Lamb 1891, 439–440). The do-what-you-can-build-now thinking became the mantra of the creators of this remarkable transportation revolution.

2

Youth, 1860–1880

TRACK STRUCTURES

The United States' overwhelming endorsement of the iron horse served as a strong impetus for the railroad's further development. All parts of the burgeoning enterprise, including managers, investors, and employees, learned that technological innovation led to a better transportation system. Moreover, anyone with a modicum of technical ability who rode trains or observed their operations might conceive of possible improvements. Since the United States was widely and correctly known as a nation of "tinkerers," individuals who were self-taught, apprentices of early civil engineers, or formally educated created and perfected a variety of components for the ever-expanding industry. Young, talented graduates of the U.S. Military Academy at West Point, New York, for example, regularly resigned their army commissions, preferring to work on matters of railroad development rather than on coastal defense projects and other military-related assignments. The challenges were more, opportunities for advancement were greater, and pay nearly always was better.

During the gestation years railroad engineers, whether West Point or college graduates, veterans of canal construction, or those who lacked much experience or formal training, quickly determined that they must focus some of their ingenuity on basic rail. Strap rail was wholly inadequate.

Heavier trains, which operated at increasingly faster speeds, caused the wrought-iron on wooden-beam structures to break apart, resulting in frequent and expensive repairs and at times costly accidents. The solution, they believed, involved *solid* wrought- or cast-iron rail. In the years before T rail became the industry standard, these pioneers would fashion their ideas in several shapes. By the Civil War era, earlier strap-rail and discarded rail designs were found only on a few lightly used branch lines and sidings and on an occasional shortline railroad, usually in an agricultural area. Yet as late as the 1870s a few impoverished carriers, like the tiny Farmers' Union Railroad in Iowa, might install primitive rail, in this case laying uncapped maple-wood stringers for track on a slightly graded dirt roadbed. Over the years the popular T rail became more substantial, often weighing as much as 70 pounds to the linear yard. This was a vast change from the earliest all-metal rails that tipped the scales at from 20 to 35 pounds.

Engineers, moreover, sought to modify the basic T design. Their goal was to strengthen and lengthen the life span of the rail. In the early 1850s the Boston & Lowell Railroad, for one, investigated the advantages of a pear-shaped rail, and other railroads, including the Baltimore & Ohio (B&O) and New-York & Erie (Erie), experimented with a compound rail. This latter style, which was tried repeatedly, consisted of two or more iron pieces that were either bolted together or interlocked much like pieces of a jigsaw puzzle. The inherent drawback centered on the failure of designers to lock these components effectively with the nuts of connecting bolts. Vibrations caused by passing trains loosened the bolts and compromised rail integrity.

On the other hand, solid, one-piece iron rail had its shortcomings. As trains grew increasingly heavier and faster, they accelerated wear, especially on curves. This necessitated extensive and hence expensive track maintenance; indeed, the latest locomotives and cars pounded iron rails to pieces. But more troubling was the tendency of iron rails to break. The ensuing wreck led to property damage, personal injuries, and even loss of life, and, of course, it usually disrupted operations. During the gestation period, when a derailment occurred it was usually merely a nuisance. Since train speeds were low (usually 20 miles per hour or less), injuries were often slight and deaths relatively rare. But if a passenger train traveling over iron rails at 35 to 40 miles per hour went "into the ditch," a disaster likely resulted.

A better technology was forthcoming. The new rail would be made of steel rather than iron. Just as Great Britain was the cradle of the railroad, so too would that nation be the birthplace of steel rails. In the mid-1850s the first pieces were rolled at the Ebb Vale Works in Wales. Although steel rails started to replace iron on lines in Britain, the dual factors of high production costs and transportation charges to the United States prevented their immediate, widespread adoption. The economic disruptions caused by the Civil War also retarded usage.

With the introduction of the Bessemer open-hearth process, an operation developed by Englishman Henry Bessemer that removed chemical impurities in a mass of molten pig iron by oxidizing them with air blown through the iron, and the return to peacetime conditions, the American steel industry took shape and grew steadily. Although the first steel rails were rolled in the mills of Chicago in 1865, Bethlehem, Pennsylvania; Birmingham, Alabama; Cleveland, Ohio; Pittsburgh, Pennsylvania; and several other places soon evolved into centers for steel-rail production. By 1873 115,000 tons of Bessemer rails were rolled, but still almost 7 tons of iron rails were manufactured for every ton of steel rails produced. With Bethlehem Steel Company beginning operations in its namesake Pennsylvania city in 1873 and the Edgar Thompson Steel Works under construction near Pittsburgh, by the end of the decade the Bessemer industry was on a solid basis. Then in the late nineteenth century steel-making changed, with manufacturers abandoning the Bessemer process for the preferred Semens-Martin open-hearth method, which employed a "regenerative furnace" that created much higher temperatures with less fuel than in the former method and in the process oxidized impurities slowly by a flame above the hearth on which the iron was placed.

Although steel rails were widely discussed, railroads were somewhat slow to embrace them, largely because of price and availability. Yet results from the earliest usage were encouraging. During 1867, for example, the Chicago, Burlington & Quincy Railroad (CB&Q) laid its first 3 miles of steel rail, installing it near Chicago where traffic was heaviest. After a two-year period the company's chief engineer reported that these experimental rails revealed little or no wear. He believed that they would last ten times longer than iron. The CB&Q then decided that it would not place on its main line any additional conventional iron rails. By 1872 the company claimed 134 miles of steel in service, and four years later the figure rose to about 400 miles. The experience at the Union Pacific Railroad, the vital link in the nation's first transcontinental rail network, paralleled those of the CB&Q. In the mid-1870s management decided that it would substitute steel for iron "as fast as it can economically be done" (Klein 1987, 425). By 1879 the Union Pacific owned 485 miles of steel line, including all curves and grades in the mountain country. As the 1880s began, the Chicago, Rock Island & Pacific Railroad (Rock Island) proudly boasted of being the only company to have an all-steel track between Chicago and the Missouri River, a physical improvement that aided it in this highly competitive rail corridor. Soon its rivals, including the CB&Q, successfully challenged the Rock Island in this field of technology replacement.

Railroads like the CB&Q, Union Pacific, and Rock Island had proof that track laid with steel rails was much more cost effective than that laid with iron rails. These carriers and others happily learned that steel rails could support heavier locomotives and rolling stock without rails breaking, thus avoiding delays, expenses, and the like. Moreover, trains made up of larger, modern cars carried freight and passengers at an appreciably lower cost per mile than did the lighter rolling stock.

The latest products of steel-rail making always seemed to be better. Manufacturers, most notably the Carnegie Steel Company (after 1901 the collosal U.S. Steel Corporation), in their quest for excellence continually revised methods of production. Over time these changes substantially altered the finished rail. Producers utilized different ores, learned various ways to manipulate the metal, and speeded up the rolling process.

At first innovation rested largely with the steel industry. But during the late nineteenth century, railroads, especially the major carriers like the B&O, Chicago & North Western, New York Central (NYC), and Pennsylvania, increasingly collaborated with these rail makers, linking their experiences more closely and systematically. After all, both producers and consumers benefited by sharing practical insights. If only one altered specifications, unfortunate results could follow. For example, when the chemistry of the steel was modified, rails might wear poorly or even fracture. On the other hand, if the shape of the rail were changed, manufacturers might be forced to try special processing techniques, ending up with expensive and generally unsatisfactory products. It made economic sense for both parties to work together for technological perfection. Rail sales constituted the vast output of steel mills and represented the largest routine capital expenditure made by carriers, large and small. (However, by the end of the nineteenth century, the Pennsylvania Railroad, the nation's largest, supposedly spent more annually for paper—forms, stationery and the like—than for steel rail, attesting to the enormous growth of its corporate bureaucracy.)

During their youth phase, railroads did not ignore the lowly, albeit essential crosstie. Although rails became the major focus of their attention when considering the track structure, improvements to this support component were considered. After the Civil War, companies turned increasingly for their crosstie needs to timber dealers rather than along-the-line farmers and landowners. It then became common for these track members to be sawed into standard lengths and thicknesses rather than hand-hewn into irregular sizes. And considerable thought was placed in the type of wood used and how to increase a tie's life span. Varieties of oak, particularly white oak, became the most popular. Not only did oak ties not rot quickly, even if left untreated, but they also held spikes well. Some crosstie producers found that they could significantly extend usage, and at reasonable costs, if the wood were treated by a zinc-tannin process, "Burnettized" as it was called, and followed with a tannin sealer.

Improved crossties offered much. Producers wanted to tap the ever-growing and profitable market. After all, the nation went on a railroad-building blitz in the 1880s. As the decade dawned, the national rail network stood at 93,267 miles; ten years later it soared to 163,587, an impressive 43 percent gain. Not only were there these additional track miles to serve, but carriers also happily discovered that increasing the number of crossties installed was a cheaper way to improve track stability than by using heavier rails. By the mid-1880s it became increasingly common on main lines to have approximately 3,500 crossties per mile rather than about 2,000 to 2,500, an increase of one-third or more. Railroads, of course, needed additional crossties, and they wished to amortize their original investments and reduce maintenance expenses with longer lasting wood.

NARROW GAUGE

Even though during the Civil War the U.S. Congress mandated standard gauge (4 feet, 8½ inches) for the first transcontinental railroad, and in the early 1870s Canadian railways opted for standard gauge, and a decade later railroads in the South and wide-gauge carriers (6 feet, 0 inches) like the Erie and the Ohio & Mississippi also had finally embraced gauge homogeneity, debate over track width still raged. During this period a narrow-gauge craze engulfed Great Britain, Europe, North America, and even Latin America. In 1871 Robert Francis Fairlie, a prominent English consulting engineer and self-proclaimed expert in slim-gauge technology, read before the British Association for the Advancement of Science a widely discussed professional paper, "On the Gauge for the Railways of the Future." In his carefully crafted presentation, Fairlie explained the financial benefits of the less-than-standard gauge railway and expected that this approach to railroad building would sweep the civilized world. It stood to be another replacement railroad technology of staggering impact. While originally favoring a 2-foot, 6-inch width, Fairlie subsequently considered 3 feet, 0 inches to be ideal for many railroad construction projects. This would become known as "Fairlie Gauge," even though there would be variations, principally the width of 3 feet, 6 inches. Soon thereafter, narrow-gauge advocates in the United States produced a rich assortment of journal articles, printed speeches, and pamphlets and even held in June 1872 a national narrow-gauge convention in St. Louis, Missouri.

Why did Fairlie and his dedicated followers embrace slim widths? The universal belief among these enthusiasts involved *economy*. A narrow-gauge railroad was presumably cheaper to build and operate than a standard-gauge counterpart. The arguments presented suggested that a narrow-gauge road in a mountainous terrain could be constructed for only about one-fifth of what a major trunk line would cost. For less difficult topography, the costs were about three-fifths. Proponents liked to argue that a double-track narrow-gauge railroad was less expensive to construct that a single-track standard-gauge line. After all, their preferred pike could have sharper curves and lesser cuts and fills because smaller equipment permitted shorter radii of curvature and allowed for narrower rights-of-way. The lower weights of narrow-gauge locomotives also made lighter, cheaper rail acceptable. And there was more. Crossties would be shorter, less lumber would be consumed in car building, and construction times for nearly all aspects of the physical plant would be reduced.

Then there was the matter of "dead" or tare weight. Fairlie repeatedly emphasized that every unnecessary inch of axle width generated ongoing

Following the Civil War the narrow-gauge craze burst upon the North American railway scene. In 1879 the Baldwin Locomotive Works of Philadelphia built this slim-width 4-4-0 general purpose locomotive for the Des Moines & Minneapolis Railroad. Later, when the Chicago & North Western Railway acquired the property and widened the line to standard gauge, this locomotive found its way to narrow-gauge trackage in Wisconsin. The small scale of narrow-gauge rolling stock is easily revealed by comparing the heights of crew members and passengers to the locomotive, baggage car, and coach. Courtesy of the author.

costs in unproductive tare weight of rolling stock. His backers wholeheartily agreed. A fleet of narrow-gauge equipment would have more favorable ratios of cargo and passenger weights to tare weight. Standard-gauge boxcars of the period, for example, had about a one-to-one ratio of dead to payload weight. This meant that for every pound of revenue cargo handled, a railroad hauled one pound of dead weight in the body of the car. But narrow-gauge advocates contended that the weight ratio was a least two to one and in the case of flatcars likely as high as three to one.

Fully believing the soundness of their logic, narrow-gauge proponents quickly started to turn into reality their dreams of low-cost rail transport. The first major American railroad to use the 3-foot, 0-inch gauge proved also to be the largest and longest lived, the Denver & Rio Grande Railway (later the Denver, Rio Grande & Western Railroad). Between 1871 and 1889 this Colorado-centered carrier built 1,635 miles of slim-width track, with the largest construction spurt occurring in the early 1880s.

What became narrow-gauge mania was closely tied to economic and political motivations. The deadly impact of the Panic of 1873, which triggered the United States' first widespread industrial depression, heightened public interest in slim-gauge technology. Farmers and small manufacturers saw high freight rates as the principal obstacle to a speedy economic recovery. And they believed that more competition in the railroad world would surely be beneficial to every customer. Therefore, the narrow gauge offered an attractive way to reform and expand the nation's railroads. It gave hope to communities bypassed by the "selfish," established carriers of a Jay Gould or a William K. Vanderbilt, and it promised cheap tariffs to the small shipper. Indeed, narrow-gauge enthusiasts argued that "the narrow-gauge railroad is the road for the people, that its cost comes within their reach and it offers to them a transportation facility superior to that offered by any other means" (Fleming 1875).

By the mid-1880s virtually every American state had at least one narrow-gauge railroad; this building boom produced approximately 12,000 miles of slim-width trackage. The construction record was impressive. In 1880 almost 25 percent of *all* railroad track miles built in the United States were narrow gauge; the following year the percentage remained high, amounting to 18 percent. Many of these roads were designed principally to haul a single commodity—coal, grain, lumber, or naval stores—and thus served as "tap lines" into the producing locales. Other projects were typical shortlines that hauled a variety of products and conducted a general passenger business.

Some entrepreneurs envisioned a regional, even interregional network of narrow-gauge lines. These projects promised inexpensive construction and operations and would usually boost "grassroots" ownership, spreading their benefits to the small investor and empowering economic reform of the powerful, entrenched standard roads. The premier effort to forge such a rail alternative to standard-gauge roads was known as the "Grand Narrow Gauge Trunk," a proposed system of several independent railroads that would stretch from Toledo, Ohio, to Laredo, Texas, with a narrow-gauge connection to Mexico City, a distance of about 1,600 miles in the United States and 800 miles in Mexico. The Grand Narrow Gauge Trunk also expected to build important appendages to Chicago and Cleveland. By the early 1880s a large portion had become reality, frequently the result of hard-working hometown sponsors who peddled stocks and bonds and received support from local taxpayers. In 1886 it was possible for a traveler to journey continuously on 1,581 miles of narrow-gauge lines between Ironton, Ohio, and Sealy, Texas. But likely no one ever did.

By the 1890s the narrow-gauge phenomenon, especially efforts to create a long-distance network, had fizzled. The inherent inability to compete with standard-gauge roads along with financial and managerial problems led to a cessation of construction. "The complete downfall of the Toledo, Cincinnati & St. Louis [Grand Narrow Gauge Trunk] system thoroughly demonstrates the impracticability of the narrow-gauge theory—particularly in the fertile and better regions of the country, when it becomes necessary to compete with standard-gauge roads," aptly editorialized the *Commercial Gazette* of Cincinnati, Ohio. "There is one thing quite certain—there will not be more narrow-gauge roads built in this portion of the Central States again soon, unless it is some very short road, and then only to fill a 'long-felt' local want" (quoted in Hilton 1990, 110). This prophecy proved correct. The vast majority of those lines that continued to operate shifted rather quickly to standard gauge or at least became double gauge by adding a third rail to accommodate standard-gauge equipment. Still, the Denver & Rio Grande Western retained large portions of its narrow-gauge system until the 1960s and two units, the Durango & Silverton and Cumbres and Toltec Scenic, survive as popular tourist operations.

Generally speaking, the attractiveness of cost was mostly an illusion. A contemporary expert argued cogently that the "projected capital savings of the narrow gauge were to only a trivial extent the consequence of the smaller gauge, but mainly were the consequence of reversion to more primitive standards that railroad engineers had already rejected" (quoted in MacGregor 2003, 85). Indeed, the first American narrow-gauge equipment designs had been inspired by light, rigid, European four-wheel cars, a concept rejected decades earlier by the industry as being unsuited for American track. Admittedly, a fleet of such rolling stock might save considerably on initial capital expenses, but these pieces often wore out quickly. Furthermore, there existed the costly problem of transshipment; the necessity to load and unload goods between narrow- and standard-gauge rolling stock at connections was hardly economical.

Yet, there were settings where narrow-gauge railroads made financial sense. A strong case can be made for California, where a sizeable network of slim-width roads appeared, especially in the greater San Francisco–Oakland area. Of great importance, the long distance from Eastern car manufacturers initially involved expensive shipping charges and considerable time delays. Several local car builders seized on this situation of geography and created viable manufacturing operations. In the late 1870s a leading car maker, Carter Brothers, developed a 28-foot, 10-ton-capacity boxcar that matched the average standard-gauge boxcar of the day and in the process enhanced the ratio of paying freight to car weight (1.5 tons of cargo to 1 ton of dead weight). Earlier this firm imaginatively produced freight car "kits" that helped to reduced labor and transportation costs. (This process involved packaging the precut and predrilled bolsters, decking, sills, siding and roofing, castings, and bolts and fasteners for the easy assembly of freight cars.) In the case of the South Pacific Coast Railroad, which in 1880 opened a 75-mile route between Alameda (Oakland), San Jose, and Santa Cruz, the problem of transshipping cargo never really developed. The road hauled lumber, grain, and other commodities directly to markets in the San Francisco Bay, explaining why during its nine years as an independent company, it was one of the most profitable railroads per mile in the state.

It was during this narrow-gauge era that some almost toylike railroads appeared. These were the 2-foot-width lines that were most popular in Maine. The wish to create more railroad for less money characterized these endeavors. The earliest of these Lilliputian roads were industrial tramways that did not handle common-carrier freight or passenger traffic, but in the mid-1870s a New Englander, George Mansfield, visited Wales and encountered the 23½-inch-wide slate-carrying Festiniog Railway. He liked what he saw. Then in 1877 the first commercial 2-footer, the 8-mile Billerica & Bedford Railroad, opened in Mansfield's native Massachusetts. The road's motive power was predictably of Bantam size and weight: the two coal-burning 0-4-4 type locomotives, which weighed less than 12 tons each, had cylinders that measured a scant 8 by 12 inches and featured 30-inch driving wheels. As Mansfield and his backers argued, the overall costs of construction were modest and total train expenses amounted to a paltry 13 cents per mile, including labor and fuel. Notwithstanding these attractive financial figures, the Billerica & Bedford failed quickly because of a lack of business, and the line was soon dismantled. "The fact that the little towns of Bedford

Builders of railroad locomotives did not always concentrate on creating larger and more muscular pieces of motive power. Following the Civil War a host of small, "dinky" steam engines were designed to meet the needs of coal mines, lumber mills, railroad contractors, and the like. About 1880, the engineer of a tiny, narrow-gauge 0-4-0 tank locomotive with its train of coal cars pauses with several associates somewhere in northeast Ohio. Courtesy of the author.

and Billerica couldn't support *any* railroad couldn't be held against the fledgling B.&B.," observed a local historian. "It had done *its* part" (Moody 1959, 54).

Although Massachusetts could claim to be the birthplace of the 2-footers, Maine became the heartland. Inspired by Mansfield's Billerica & Bedford, local promoters in the heavily timbered section of southwestern Maine launched the 18-mile Sandy River Railroad that opened in 1879, becoming the harbinger of several other companies that collectively laid approximately 200 miles of track. As expected, these roads hauled mostly lumber products (and, in the case of the 6-mile Monson Railroad, principally slate). A summer resort passenger business also developed on the largest carriers. This 2-foot kingdom proved remarkably enduring: the Bridgton & Saco River Railroad operated until 1941 and the Monson lasted two years longer. While hardly a national phenomenon, the logic of Mansfield that a practical railway of only 2-foot gauge could be operated had some merit. In reality these 2-footers, "the smallest of the small," became a form of "niche" replacement technology.

Although the 2-footers were never changed to standard gauge, hundreds of miles of viable narrow-gauge lines were widened. The process was relatively simple for the track. Work crews smoothed the ties where the rail was to be shifted and placed along the worksite a row of spikes. Then it was only necessary to loosen the rails on one side and move them over to the correct position. Yet altering rolling stock was more complicated as well as time-consuming and expensive. Personnel needed to jack up the rolling stock, remove the trucks, and then disassemble them. A wheel lathe was employed to cut away metal from the axle, and a mighty wheel press moved the wheels to the correct gauge width. Laborers then reassembled the trucks and placed them on the equipment. But once completed, regauging was "forever"; no railroad ever reverted to a previous nonstandard width.

BRIDGES AND TUNNELS

No matter the gauge selected, railroad builders needed to bridge waterways and occasionally to burrow through natural barriers. While tunnels were seldom required, except when roads crossed mountainous terrain, bridges were ubiquitous. From the dawn of the Railway Age, engineers and crews labored to span both large and small bodies of water.

For centuries the art of bridge building had been refined until craftsmen knew just how these spans should be shaped. One of the oldest bridge forms was the masonry arch, dating from the earliest recorded periods of civilization. The builders of the B&O, Western Rail Road of Massachusetts (Western), and other important antebellum roads opted for the masonry arch for their first major spans. Generally speaking, the engineering and technologies embraced were not particularly novel. Workers built the arches in place, cutting, laying, and mortaring the large stone blocks, and in the process erected temporary timber "centering" to brace each arch until they finished their stone work. Roads like the B&O and the Western often enjoyed convenient access to good supplies of building materials: stone was quarried as near as possible to the bridge site and useable timber was commonly found in close proximity.

The results of the skills and hard work of these long-forgotten artisans have stood the test of time. The Starrucca Viaduct, which carried the Erie Railroad over the deep valley of the Starrucca Creek in northern Pennsylvania, for example, not only has been recognized for its grace and beauty but also continues to handle heavy trains more than 150 years after its completion. Fashioned of enduring blue stone, its eighteen tall slender arches rise 110 feet above the sluggish stream at its base. It is not a particularly long structure, 1,200 feet from end to end, but it is unquestionably stout and built for the ages. The cost of more than $300,000, which made it the most expensive bridge ever built anywhere in the world up to that time (1848), turned out to be an excellent long-term investment.

Yet railroad builders found shortcomings in stone-arch bridges. As indicated by the cost for the Starrucca Viaduct, these structures were expensive. Moreover, they required the services of skilled craftsmen and took months, even years, to complete. Throughout much of the nineteenth century the American railroad network grew by leaps and bounds, and companies needed cheap and easily constructed bridges. Fortunately wood was plentiful and inexpensive, allowing for short-term spans that were hardly erected for future generations. No wonder foreign visitors remarked on the popularity of wooden bridges, in part because of their ease of construction and low price tags. "The *bridges* on this line [Utica & Schenectady Rail Road] are all of wood, and with a few exceptions all have stone abutments," observed German railroad expert Franz Anton Ritter von Gerstner in the early 1840s. "On the smaller bridges, the wooden superstructure usually consists of simple truss work. Costs for this were $5 per running foot on bridges with a 40 foot span, $6.50 on those with a 60 foot span. The large bridge across the Mohawk River at Schenectady consists of 5 arches, each with a span of 145 feet. The superstructure is built entirely of planks 6 inches wide and 3 inches thick, and total costs amounted to $35,000" (Gamst 1997, 162). In von Gerstner's eyes these American roads were usually primitive affairs, built quickly and cheaply. Later, if financial conditions and traffic warranted, carriers like the Utica & Schenectady assembled better bridges with metal components, or they opted for earthen fills with culverts to replace the original construction.

During the Civil War, American bridge-building ingenuity was vividly demonstrated. Herman Haupt, most famously, repeatedly showed his engineering genius. This graduate of West Point and author of the influential *General Theory of Bridge Construction* (1851) determined how replacement spans, necessitated by wartime destruction, could be built in record time to carry heavy military loads, even though, as President Abraham Lincoln quipped, they appeared to have "nothing in [them] but beanpoles and cornstalks." Yet like all wooden structures, these bridges were vulnerable to fires. Not surprisingly, railroads routinely hired "track walkers" to make inspections, usually after the passing of a wood- or coal-burning train that may have drenched the structure with sparks and red-hot cinders.

Haupt was wholly familiar with the bridge type most widely used for much of the mid-nineteenth century, the improved timber truss design developed in the late 1830s by New Englander William Howe. Although used for public wagon bridges, the Howe truss became a favorite among railroad builders because of its strength, durability, ease of fabrication, and low cost. The Howe truss typically combined heavy top and bottom horizontal members or chords and numerous wooden diagonal pieces with vertical tension members of wrought iron. It did not take long before the all-metal (and largely fireproof) Howe truss bridge appeared, the first being a 34-foot cast- and wrought-iron span that the Philadelphia & Reading Railroad constructed in 1845.

As railroad bridges became stronger and more numerous, civil engineers paid greater attention to their development. After all, they needed to erect spans that could safely manage increasing locomotive weights and operating speeds. The design of these bridges remained largely empirical in nature, an estimation of the uniform live load on the structure. "The science of bridge proportioning was yet undeveloped," observed Theodore Cooper in his widely praised *American Railroad Bridges*, published in 1889 by the American Society of Civil Engineers. "The best that the engineer could do, was to make the bridges stronger than heretofore, solely on the facts brought out by past experiences" (quoted in Middleton 1999, 3). Cooper, though, made a major contribution when he developed a widely used classification system of locomotive and train loading for bridges that still bears his name.

As with most other aspects of railroad technology, by the post–Civil War era there were signs that *scientific* bridge analysis and design had fully begun. Observers rightly commented on the remarkable progress of railroad-bridge building. In 1870, for instance, the greatest span designed by German-born and -educated Albert Fink was completed for the Pennsylvania Railroad to cross the powerful Ohio River at Louisville, Kentucky. This creative civil engineer used a novel iron-truss type for the principal river channel spans of 370 and 400 feet. What became the longest truss bridge would be used for the next forty-seven years. Fink's triumphal approach to large bridge building would be widely copied.

Thirteen years later trains of the Michigan Central Railroad rumbled high above the Niagara River over one of the most important early cantilever bridges. Made of wrought iron and recently introduced steel, the anchor arms, located at each end of the structure, measured about 200 feet in length, and cantilever arms of 175 feet (that projected toward each other) and a 120-foot "suspended," or connecting section, gave the bridge a clear span of nearly 500 feet. Like steel rails, the steel components soon revealed that this replacement metal possessed much greater strength, toughness, and resistance to wear than either cast or wrought iron.

The nearby great gorge of the Niagara was the site of the amazing Grand Trunk Railway suspension bridge, designed by another German immigrant, John Augustus Roebling, famous for his Brooklyn Bridge over the East River in New York. By using specially designed and fabricated wire rope or cable, it was possible to devise a more than 800-foot span that straddled two massive limestone towers. This structure claimed the distinction of being the only railroad suspension bridge ever built in North America. Although it was generally believed that a suspension bridge could not be constructed to be rigid enough for railroad traffic, Roebling succeeded in his highly experimental approach to bridge building. A network of deep stiffening trusses and wire rope stays maintained stability. When this engineering landmark opened for revenue traffic in 1855, there were, according to its creator, "No vibrations whatever." However, by the 1890s, even with several major modifications and regular maintenance, the Roebling suspension bridge needed to be replaced. Freight and passenger train volume had increased substantially, and the railroad required a span that could accommodate double tracking. It also required a structure that could handle speeds greater than the 5 mph restriction bridge engineers had imposed, and it needed to safeguard always increasing train weights as well.

By the 1880s the steel age in railroad bridge construction had arrived, and during the following decade it came fully of age. When building its Kansas City extension in the late 1870s, the Chicago & Alton Railroad claimed to have been the first carrier to install an *all-steel* bridge, and perhaps it was. The company needed to cross the wide and meandering Missouri River at Glasgow, Missouri, and initially it decided to use five long iron spans. The bridge supervisor, William Sooy Smith, who also chaired the American Society of Civil Engineers' Committee on Bridges, learned of a process that inventor A. T. Hay had recently perfected, namely, a steel alloy that was superior to iron. Smith became a believer and risked the recommendation to the American Bridge Company that the Hay steel alloy be used. Although iron-bridge builders predicted disaster, workmen erected the temporary wooden "false work," or scaffolding, installed iron approaches, and then one by one positioned the steel spans. Even though the pressure of a buildup of river ice caused the false work to shift, dislodging a span, work progressed steadily, allowing the structure to receive successfully its first train on April 9, 1879.

While design and building materials were at the heart of the evolution of the railroad bridge, other techniques also contributed mightily to engineering advances. Without doubt, improvements in the construction of foundations played a significant part in bettering spans. Early on engineers devised the cofferdam, a watertight, albeit temporary structure that was placed in a river or lake to keep water out of an enclosed area that had then been pumped dry. This protected space permitted workers to construct foundations below the normal level of the water. Although cofferdams worked well in relatively shallow bodies of water, locations in deep water caused major challenges, largely due to mounting water pressure and pumping difficulties. Fortunately, an alternative emerged, the pneumatic caisson. In what might be considered an upside-down box from which water was excluded by air pressure, excavation and construction work could take place. Like the cofferdam, the caisson acted as a dam, keeping out mud and water. In 1868 James Eads first used the pneumatic caisson process during the construction of his remarkable railroad and public wagon bridge across the Mississippi River at St. Louis. Although the caisson was dangerous for workers early on because of "caisson disease," which was the "bends" or rapid decompression, Eads soon devised the first depressurization apparatus, a predecessor to the modern hyperbaric chamber, that slowly acclimated the affected individuals to atmospheric pressure. These refinements to Eads's pathbreaking work made possible numerous large bridges, both for trains and private vehicles.

The nation's growing cadre of civil engineers did more than build railroad bridges over water; they also installed structures for them below ground. Although tunneling has been ongoing since ancient times, tasks were often herculean; the use of metal wedges and heat were backbreaking and time consuming. But in the seventeenth century a monumental breakthrough in tunneling techniques occurred when German coal miners began to drill and blast with gunpowder. Since railroad builders sought to create a roadbed that was as level as possible, hills, ridges, and mountains offered considerable challenges. Tunnels, of course, offered a practical solution to these natural barriers. Between 1826 and 1830 the first railroad tunnels appeared in France and Great Britain. And it would not be long before the first railroad tunnel was driven in North America: the 901-foot bore through a slate hill for the Allegheny Portage Rail Road near Johnstown, Pennsylvania, which laborers fashioned during 1831–1832. More tunnels followed, and by midcentury nearly fifty of these engineering and construction triumphs dotted the American landscape.

These early railroad tunnels were built of similar construction. Tunnelers drilled holes in the rock walls with hammers and chisels, and then filled those cavities with black powder for the blasting process. At times they not only bored from the two entrances but also sank one or more shafts from the top to facilitate construction. Teams of horses or mules pulling small wagons or rail cars removed the debris or "muck." Workers frequently used this discarded material for leveling the approaches to the tunnel. Depending on the stability of the excavation, laborers might install a wooden or masonry lining before a track gang laid the rail. This finishing step did not always occur. In the late 1830s, for example, when workers built the New York & Harlem Railroad in the greater New York City area, they created a 600-foot tunnel that was 24 feet wide and 22 feet high that, according to an early visitor, "was blasted out of rock consisting of quartz and hornblend and hence required no masonry work"(Gamst 1997, 246).

With the ongoing need for tunnels, especially longer bores, and the slow and costly nature of construction, technological improvements became essential. And they would occur in the 1860s and after. Compressed air-rock drills, dynamite, and nitroglycerin made tunneling faster and less expensive. Yet these betterments did not mean that dangers were dramatically lessened. More so than bridge building, tunneling remained hazardous. The materials to be bored meant difficult and life-threatening work; "unconstituted rock" or "rotten sandstone," for example, made for a treacherous undertaking. In 1878 the *Mining and Scientific Press* described a railroad tunnel undertaking near Santa Cruz, California:

> The tunnel is arched at the top and its sides incline slightly from the spring of the arch to the floor. The distance from the crown to the floor is 16 feet three inches. The span of the arch is 13 feet. The width of the floor is 11 feet four inches. The timber is all 10 by 12 inch stuff. The arch is made in six pieces. Each is three feet 10 inches on the top and three feet four inches on the bottom. A flat rod of iron runs through each of the these joints, connecting each set of timbers firmly with the adjacent ones. Behind the timbers, four by four-inch lagging is placed all around throughout almost the entire length of the tunnel. This is necessitated by the shifting nature of the ground. (Quoted in MacGregor 2003, 532–533)

These California railroad-building endeavors experienced other troubles, including a nasty explosion. Reported a local newspaper,

> On this occasion the gas, and the oil that had gathered at the bottom of the tunnel, both caught fire, the sheet of flame being fearful to contemplate. At the time of the explosion, [a tunneler] was at work at the face of the tunnel, and immediately fell down, lying on this stomach and placing his hands by the sides of his face. He got up twice and tried to run, but the heat was so intense that he had to lie down and crawl on his hands and knees. Fire above him, fire around him, and a round body of fire from one end of the tunnel to the other. . . . Half way down the tunnel were the two Chinamen who are supposed to be dead. They were in charge of the cars, two in number, which were thrown from the track, and the mule hauling them was carried for a distance of sixty feet towards the mouth of the tunnel. A brace, two hundred feet from the mouth of the tunnel and eight feet from the mouth of the tunnel and eight feet above the bottom, was carried seventy feet beyond the mouth of the tunnel . . . the blacksmith shop, half a mile from the place of explosion, was blown to atoms. . . . The report was so great that people outside of the tunnel thought the [powder] magazine had exploded. (Quoted in MacGregor 2003, 543)

While dangers always existed, engineers developed an important building technique that made the construction of hard-rock tunnels somewhat easier. Known as the "heading and bench" method, workers would drill a heading (or drift) within the cross section of the tunnel. This involved creating a heading at the top that extended the full width of the bore. Later tunnelers excavated the lower portion (or bench). For long tunnels as many as a half dozen or more drifts or headings would be used at one time, facilitating swifter completion of the project.

No matter the length or the approach employed, a repeated danger that tunnelers encountered involved underground pockets of water, at times in enormous quantities. In 1853 Claudius Crozet, a French-trained pioneer in railroad construction techniques, reported the water problems that workers encountered as they bored the troublesome, albeit rather short Brooksville Tunnel for the Virginia Central Railroad:

> We were suddenly taken by surprise by the eruption of a large vein of water, for which we were obliged to take hands from the work and set them to pumping. . . . This circumstance has been repeated several times during the year until the body of water we now have to keep down amounts to no less than probably one and a half hogsheads per minute. (Quoted in Warden 1973, 51)

Finally, Crozet subdued the underground springs with a creative 1,800-foot long and 3⅛-inch-thick siphon pipe laid down the mountain slope. This was said to be the longest siphon ever built up to that time.

Occasionally, railroads in the nineteenth century needed tunnels that ran underwater. The combination of soft or fluid materials found beneath waterways together with hydraulic pressure created a significantly different building environment, requiring that traditional construction methods be modified. Yet as the Railway Age dawned, examples of subaqueous tunneling occurred. English engineer Marc Isambard Brunel supervised the construction of one of the first major underwater tunnels, a bore under the Thames River in London. Between 1825 and 1843, workers used Brunel's specially designed cast-iron "shield" that supported the water-bearing clay of the river bed as the tunneling progressed. Tunnelers then installed a permanent brick lining behind the shield. The innovative Brunel shield concept became the common way to create these expensive and time-consuming passages until the 1870s when a Californian, DeWitt Clinton Haskin, employed the use of compressed air caissons. His patented method substituted air pressure for the shield to keep water out of the tunnel until a permanent lining could be installed.

DEPOTS AND TERMINALS

While tunnels were relatively unusual, depots and terminals appeared widely. The most numerous station structures were those that served thousands of smaller communities. By the Youth Period, railroads had largely formalized the design and construction techniques for such essential buildings. When carriers pushed ahead of population, which commonly occurred in the 1870s and 1880s, they developed a simple prefabricated portable depot. These were small enough to be easy to assemble and to transport on a flatcar to wherever a depot was needed. But as communities grew, larger replacement structures became mandatory. Designing small, attractive, low-cost depots that would please town boosters and satisfy local business requirements became an ongoing corporate concern, usually in urban centers of the East and Midwest.

A fundamental issue, never completely resolved, was the extent to which a railroad could afford to erect architecturally attractive structures. *Engineering News*, for one, argued that companies had gone too far in making their stations architectural gems. This publication felt that while "the passenger stations of American railways have within recent years shown a marked advance in their . . . design," practical improvements were required. Too frequently, meeting the needs of the traveling public was "not given sufficient attention," or these requirements were "subordinated to architectural treatment." In short, railroads had become so responsive to the desire for architecturally attractive stations that companies ignored opinions of operating officers on design. Depot planning for the *Engineering News* was too important to "be entrusted entirely to an architect, even if he be one of the regular officers of the road" (quoted in Grant and Bohi 1978, 18–19).

Railway Age, the leading industry trade magazine, however, took a different position. Depot design did not "always receive the professional treatment it deserves." Railroads had an obligation to the community that went "beyond the ordinary selfish requirements of the owner." It was not enough that a station be functional; it must also be attractive. "In this way . . . railroads could do much in the way of elevating the taste of a community" (quoted in Grant and Bohi 1978, 19).

Another *Railway Age* issue discussed the closely related industry concern about projecting a positive public image. Describing how the Michigan Central Railroad established a program to "embellish" its station grounds, the publication carefully explained why the company thought the cost justified. It was believed that an attractively kept depot area served as a fine advertisement for the road. Working hard to make a town's gateway beautiful could graphically show "that a railroad may be something other than a mere sordid, money-making machine it is often credited with being" (quoted in Grant and Bohi 1978, 19). Some of the pioneers and innovators in landscape architecture won assignments from railroads.

Discussion of depot design was not limited to trade periodicals. In 1893 a book appeared with the ambitious title *Building and Structures of American Railroads: A Reference Book for Railroad Managers, Superintendents, Master Mechanics, Engineers, Architects, and Students.* Written by Walter Gilman Berg, chief engineer of the Lehigh Valley Railroad and an internationally recognized authority on railway building and maintenance-of-way matters, this work devoted nearly 100 pages to a treatise on station design. One chapter focused on the small-town, "country," or "combination" depot. Although primarily a survey of station plans used by different railroads, Berg's book generalized about such buildings. "[They should be] used on railroads at [locations] of minor importance," he contended, "where the amount of freight or the volume of the passenger business does not warrant the construction of a separate freight-house or a separate passenger depot." Diversity of design, he felt, was inevitable because nonurban depots had to meet a multitude of specialized needs. As Berg said, "A large number of variations exist in combination-depot designs, according to the necessity of providing for and the relative importance given to the freight service, passenger business, baggage, express, telegraph, etc., and whether and how much room for dwelling purposes has to be reserved [some depots provided living space for the agent and his family]" (quoted in Grant and Bohi 1978, 20–21). Dividing a building into three basic parts made considerable sense; namely, a public waiting room, a freight and express storage section, and an agent's office, the latter usually located in the center of the building and commonly featuring a trackside bay window. Yet, Berg's survey of combination depot drawings demonstrates that most carriers, even by the 1890s, employed *standardized* plans as much as possible. This is hardly surprising; by the post–Civil War era, the quest for standardization had become a powerful part of American industry and American life, and ranged from standardized machine components to standardized rules for baseball.

Not part of the standardized structure policies of youthful railroads was the urban passenger station. By the Gilded Age, usually defined as the decades of the 1870s and 1880s, railroads needed some massive specially designed structures, and cities likewise desired to have terminals that were both convenient and beautiful. Since the costs were high, railroads often joined forces to share costs, creating *union* depot companies that could plan, build, and operate such facilities. Inevitably an architectural firm, rather than a railroad's engineering or "bridge and building" departments, drew the design and supervised construction work.

Construction of these wonderful monuments to the transportation boom followed current building practices. As skyscrapers began to appear during the Gilded Age, railroads sought similar, massive structures. Take the case of the gigantic Romanesque-style St. Louis Union Station, which in 1895 opened in this Missouri railroad Mecca. Gangs of skilled artisans and semi- and unskilled laborers, under the watchful eyes of the architect and contractors, assembled this noble building. A workforce that numbered in the hundreds cleared the multiacre site, installed concrete footers, which were partly reinforced with steel, and fashioned rock foundation walls that they finished with granite from the top of the grade level. Workers subsequently faced with Indiana limestone and backed with courses of durable brick the principal east and north facades of the depot, technically known as the "headhouse." On the south and west walls they laid thousands of buff-colored Roman bricks below the level of the attached midway roof of glass and iron (which linked the headhouse to the trainshed) and gray brick above, and used for the lower floors noncombustible materials to meet contemporary standards of fireproof construction. On the upper floors these crews installed heavy timber structural members that were specially prepared to have "slow-burn" qualities. Finally, laborers covered the roof with Spanish tiles in a gray color to match the walls below, according to architect Theodore Link "on the theory that a monochrome will aid in attaining an effect of loftiness" (quoted in Grant et al. 1994, 70). As with other late-nineteenth-century buildings, the contractors selected modern mechanical services, including the latest and most fashionable plumbing fixtures. The signature feature of St. Louis Union Station was its tower, which soared 230 feet above the track level. The tower not only featured a decorative, albeit useful clock on each side but also contained an inlet for the ventilating system, another state-of-the-art aspect of the complex's mechanical system.

While the headhouse was both utilitarian and attractive, the station facility also contained the massive trainshed. Designed by the engineer George Pegram, this steel structure, with its vast curved roof supported on large arched trusses or ribs, measured 700 feet in length and 606 feet in width and covered thirty-two tracks. During this period, few structures had the sheer architectural drama of the great balloon trainshed, and this St. Louis structure was the biggest of them all. The most modern construction features were employed, including devices that allowed for expansion and contraction due to temperature changes, often a severe problem in all-metal buildings like a trainshed. A system of slotted holes with bolts for north and south movement and rollers under the trusses for east and west motion solved potential problems of any major thermal-induced shifts.

LOCOMOTIVES

Just as colossal union passenger terminals like St. Louis Union Station appeared in the principal railroad centers, so too did steam locomotives of growing size and strength. By the Gilded Age the "steam kettles" of the antebellum era were becoming a thing of the past. In 1845 the typical weight of a locomotive was 18 tons, tractive effort (similar to drawbar pull) was 4,500 pounds, and horsepower amounted to 240. A decade later the average weight for a locomotive was 25 tons, tractive effort stood at 7,010 pounds, and horsepower 375. By the close of the Civil War, engine weight had increased to 30 tons, tractive effort to 9,830, and horsepower to 520. By 1875 weight was 32 tons, tractive effort 10,500 pounds, and horsepower 560.

Although American standard locomotives (4-4-0) continued to be used and even improved for both freight and passenger operations, larger, more powerful, and more specialized motive power came into widespread use. Locomotives now had additional wheels and different wheel configurations (see Appendix). Just as the 4-4-0 was once the industry workhorse, the "Ten-wheeler" or 4-6-0, introduced about midcentury, often became its replacement. Engines designed for passenger service, whether a 4-4-0 or 4-6-0, usually featured large-diameter driving wheels that allowed high-speed operation. On the other hand, main line freight locomotives and yard switchers commonly possessed small-diameter drivers that produced greater pulling force but moved at slower speeds. In fact, the 2-6-0 ("Mogul") and 2-8-0 ("Consolidation") frequently handled over-the-road freight trains, and the 0-6-0 ("Six-wheel switcher") and 0-8-0 ("Eight-wheel switcher") usually shunted cars around yards and sidings.

The perfection of more useful locomotives involved much more than their wheel arrangements. Design engineers not only opted for steel instead of wrought iron for boiler construction that allowed for much higher boiler pressure but also experimented with the size and configuration of these boiler shells. This occurred not only because of the desire to increase power but also because lump coal replaced cord wood as the principal fuel. (For some railroads in the Mid-Atlantic region, this meant the use of anthracite or "stone" coal that often came from company-owned mines in northeastern Pennsylvania.) The Belpaire boiler, designed by Alfred Jules

In 1908 James J. Hill (individual with hat on right), founder of what became the mighty Great Northern Railway, stands beside the *William Crooks*, the first piece of motive power on the St. Paul & Pacific Railroad, the core of the Great Northern. This classic American 4-4-0 handled both freight and passenger train assignments. Courtesy of the author.

Belpaire, created considerable interest. It featured longitudinal bulges on each side near the top that resulted in a flat upper surface, a shape that produced enhanced structural strength and provided a greater surface area for heat transfer. Its high price tag, however, retarded its usage on most railroads.

Builders also sought to perfect fireboxes, especially for locomotives that selected anthracite coal because of its availability and cheapness. In 1877 John E. Wooten, a high-ranking mechanical official of the Philadelphia & Reading Railroad, introduced a firebox that bore his name. His invention enhanced steaming capacity and allowed the engine to burn anthracite. Larger fireboxes, with their expanded grate area, better handled the slower burning hard coal.

Since locomotives needed to transfer water from the attached tender to the engine boiler, early in the development process considerable attention focused on feedwater pumps. By the 1850s mechanics believed that the most successful device was the injector. This betterment was one of the few railroad innovations that developed from a *scientific* background. A French engineer and inventor, Henri J. Giffard, using Venturi's principle, one of the great laws of fluid mechanics, found a way to force water into a boiler by using steam that had passed through various nozzles and tubes resulting in higher velocity and greater pressure. Unlike feedwater pumps, the compact injector system lacked moving parts; it preheated the water and operated while the locomotive was standing. Although some technical problems existed, by the 1880s developers had made the injector wholly workable. It became standard on all locomotives, even the smallest and cheapest.

Locomotive valve gears also received repeated refinement. During the Youth Period a number of types were developed. The most successful valve-gear system was the Walschaerts design, the brainchild of Belgian inventor Egide Walschaerts, which dates from the 1840s. But it took more than thirty years before a technology transfer between Europe and North America occurred. In 1875 the Mason Machine Works, located in Taunton, Massachusetts, began to employ Walschaerts's patented value gears on its locomotives. These reciprocating value gears, perhaps the most complicated assemblies found on a steam locomotive, handled the admittance of steam into the power cylinder and exhausted it at the proper intervals during the revolution of the drivers. Later, a device with fewer parts, the Baker gear, developed and manufactured by a locomotive supplier, the Pilliod Company of Swanton, Ohio, competed effectively with Walschaerts's creation.

Furthermore, the basic throttle came under revision. Designed to regulate the supply of steam fed to the cylinder's value, the throttle valve caused no particular problems as long as steam pressure remained low, but in the Youth Period that changed as locomotives became larger and more powerful. Not only did mechanics experiment with the proper placement of the throttle value, but by the 1870s they also believed that installation of a poppet value would improve operations, although a prototype dated from the late 1840s, suggesting that an invention and a practical application were often far from being simultaneous. With this betterment the engineer, by means of a ratchet and latch, could set the throttle in any position desired. There was also an important safety feature: the throttle could be locked shut.

Toward the close of the Youth Period railroad mechanics labored at increasing the efficiency of the steam locomotive. In the mid-1870s the price of coal amounted to about 30 percent of the overall cost of operations; by the early 1890s fuel costs had risen to more than 40 percent, making fuel the *most* expensive factor in running a locomotive. A practical solution involved using steam, and the result was the "compound" locomotive. While the first compound engines appeared soon after the Civil War, the common application did not occur until 1889 when Samuel Vaulcain, an employee of the Baldwin Locomotive Works, the industry's dominate manufacturer, produced a locomotive copied after a British model.

At times known as a "double expansion" locomotive, the compound realized more power from its steam than did a simple locomotive. The typical steam engine of 1890 had its steam heated to a temperature of about 450 degrees Fahrenheit, thus achieving around the firebox a pressure of about 200 pounds per square inch. Then by the time the steam was drawn into the power cylinders, it had dropped to about 220 degrees Fahrenheit and about 85 to 90 pounds per square inch. The compound engine handled this decline by first employing the steam in a *high-pressure* cylinder and then by reusing it in a *low-pressure* cylinder. Because of the higher pressure, the first cylinder needed to be smaller than the second.

By the time compound locomotives appeared in sizable numbers, locomotive builders had developed more efficient, modern construction methods. Before 1880 most locomotives were fairly small, with the eight-wheel Americans (4-4-0) being the most common. Weights rarely exceeded 40 tons. Assemblers did not need big overhead cranes; blocks, jacks, hand-powered swing cranes, and human muscles allowed workers to handle shop-floor tasks. A decade later this was no longer the case. The growth of locomotive size required more than strong backs and clever minds. It now became essential to employ large power cranes and to create roomy overhead clearances to manage production requirements.

Since the steam locomotive contained a vast array of parts and mechanisms, constant changes were occurring. Some were patented and others not. But with the notable exception of injectors, most improvements during the Gestation and Youth periods came about through the efforts of *practical* mechanics effectively solving mechanical problems. "These men, if not suspicious of 'fancy theories,'" observed John H. White Jr., the foremost scholar of the American locomotive, "were in the main unschooled in the abstract principles of engineering and the laws of physics" (White 1997, 128).

ROLLING STOCK: THE FREIGHT CAR

From the dawn of railroading, shipments of freight were the raison d'être for this new form of transportation. Just as the steam locomotive evolved steadily during the Youth Period, so did the lowly freight car. After all, the nation grew dramatically after the Civil War and with urbanization and industrialization came the need for larger and specialized freight equipment. An indication of the growth of major products that railroads hauled are revealed in these figures of goods shipped for 1880 and 1890 (listed in millions of tons):

Commodity	1880	1890
Coal	89.6	184.3
Merchandise	58.2	NA
Grain	42.0	49.4
Lumber	25.4	52.8
Iron and steel	11.6	21.0
Livestock	10.8	15.4
Petroleum	7.7	8.7
Flour	7.4	11.2
Dress meat	NA	6.1
Cotton	3.9	4.9

During the Youth Period the trackside observer readily noticed the increasing size of the boxcar, especially when modern cars were placed next to older pieces of equipment. Capacity grew from 10 to 15, 20, 30, and finally 50 tons and cubic volume roughly doubled, increasing from approximately 1,000 to 2,000 cubic feet. Moreover, tare weight (weight-to-capacity) dropped by nearly 40 percent. Although cars remained built principally of wood construction, cast- and wrought-iron parts had mostly given way to malleable iron and steel.

Although the boxcar was the most popular piece of freight rolling stock, speciality equipment became commonplace. Trains usually contained tank cars, coal cars, flatcars, gondola cars, refrigerator cars, and stock cars. The latter, for example, might include live poultry cars or even exotic horse cars.

While each type of freight car had its own evolution, the refrigerator, or "reefer" car, experienced its birth, youth, and early maturity between 1850 and 1900. Even though this may not be the "typical" piece of rolling stock, its development reflects trends in the car-building world. As is commonly the case, no one person likely "invented" the reefer. Rather, its gestation probably came from several individuals working independently of each other before the Civil War. It is known that in 1851 an employee of the Ogdensburg & Lake Champlain Railroad (O&LC), Jonas Wilder, convinced the company's master car builder to remodel several boxcars with charcoal insulated walls in order to send locally produced butter to the Boston market. These refitted cars were then loaded with butter and ice, and more ice was added en route. Not long thereafter Wilder, who had left the O&LC for the nearby Rutland & Washington Railroad, continued his interest in refrigerator cars for shipments of butter, cheese, and meat and even hit upon the idea of installing heaters for wintertime transport of potatoes.

Less than a decade later a New Yorker, a Mr. Lyman, developed his vision of a reefer. His cars were described as being "zinc lined with double walls filled with cork. Ice bunkers, two feet deep and four feet long were placed at either end [of the car], and an inclined drip floor and pipes carried away the water from the melting ice." Also part of Lyman cars were "roof hatches . . . above the ice bunkers," designed for convenient loading of block ice (White 1986, 31).

As would be expected, betterments were not long in coming. In the 1870s Joel Tiffany, a Chicago lawyer turned refrigerator-car maker, concocted a modestly successful car that featured a full-length, overhead ice bunker. About the same time James Wickes, another inventor, devised a sheet-metal ice bunker that became widely used. And another contemporary of Tiffany and Wickes, John Ayers, introduced the "Ayers's Rubber Refrigerator Car" that featured a thin sheet of India rubber to block the flow of outside air that penetrated the cracks found in the wooden car body. Ayers unquestionably had made a major contribution to the art of reefer-car making.

At times inventors seemingly wasted their time. What these individuals did was to devise highly ingenious improvements and gadgets. Floor fans, for example, were not particularly practical. Railroad operating personnel wanted reliable and *simple* refrigerator cars that would not break down or need extensive attention. In order to be successful, even the dimmest trainman needed to be able to handle the equipment.

ROLLING STOCK: THE PASSENGER CAR

During the Youth period, passenger train riders commonly encountered greatly improved rolling stock. They usually found longer car body lengths and increased seating capacities, and probably realized that the equipment weighed considerable more. Indeed, if a comparison were made between a typical passenger coach operated by the Chicago & North Western for 1885 with a car used by the Baltimore & Ohio fifty years earlier, the vital statistics are as follows: body length, 55 feet versus 30 feet; seating, sixty-four versus fifty; and weight, 30 tons versus 6 tons. Moreover, the inside surroundings were more inviting. In the mid-1880s the *National Car Builder*, for example, described a newly completed Erie coach as having a "pleasant, cheerful and attractive interior." And it added, "The seat-arms are of ash, and are strong, shapely and handsome. The seats slide upon the frames so as to be lower at the back than at the front edges. The roof curves have been carefully studied with a view to harmony between those of the lower and upper roof." Concluded the publication, "The raised roof, in consequence of the thickness of its sides being only 1½ inches, is much lighter than usual, without any sacrifice of strength" (quoted in White 1978, 95).

Undeniably, then, better construction technologies became part of the passenger-car story and car designs improved. Following the Civil War Ezra Miller, a civil engineer, for one, introduced the elevated platform, a relatively simple alternation that enhanced safety by making for a more desirable way of joining cars. Miller, too, called for the use of a strong buffer beam, namely, the end cross timber of the platform, and the use of iron rods to strengthen the overall car body. Yet the mostly metal passenger car would not really appear until the early part of the twentieth century. After all, hardwoods were plentiful and not particularly expensive, and the lumber could easily be shaped and crafted by master carpenters. Moreover, in the late 1880s the invention of the covered vestibule reduced the violent jerking motion of platforms and made it easier for passengers to move safely between cars.

Yet, it would be farfetched to imply that all passenger rolling stock was luxurious or even comfortable. Major railroads continued to use cars that were built in the 1840s and 1850s, especially on locals and in branch line service, and old equipment remained common on shortlines, particularly those that operated in rural locales. And for the humblest or lowest-class operations carriers might own new, albeit cheaply constructed and inexpensively furnished coaches. In the latter part of the nineteenth century when Robert Louis Stevenson, the Scottish writer, made an excursion between the coasts of the United States, he found much to dislike about the "emigrant cars" that the Union Pacific assigned to carry this clientele. As he wrote in *The Amateur Emigrant*,

> Those [cars] designed for emigrants on the Union Pacific are only remarkable for their extreme plainness, nothing but wood entering in any part into their construction, and for the usual inefficacy of the lamps, which often went out and shed but a dying glimmer even while they burned. The benches are too short for anything but a young child. Where there is scarce elbow-room for two to sit, there will not be space enough for one to lie. Hence the company's servants, have conceived a plan for the better accommodation of travelers. They prevail on every two to chum together. To each of the chums they sell a board and three square cushions stuffed with straw and covered with thin cotton. The benches can be made to face each other in pairs, for the backs are reversible. On the approach of night the boards are laid from bench to bench, making a couch wide enough for two, and long enough for a man of the middle height; and the chums lie down side by side upon the cushions with the

> head to the conductor's van and the feet to the engine. When the train is full, of course, this plan is impossible, for there must not be more than one to every bench, neither can it be carried out unless the chums agree. (Quoted in Grant 1990, 53)

But well-to-do patrons did not confront the lowly emigrant car or older, almost historic rolling stock. When in 1877 Florence Leslie, the wealthy wife of the publisher of *Frank Leslie's Illustrated Newspaper*, made a transcontinental journey, she found considerable pleasure and excitement with her modern railroad travel accommodations. The first major leg of her trip from New York City involved the use of a Wagner-built sleeping car and the second segment from Chicago westward featured a "Pullman Hotel Car," which was a combination sleeper and drawing-room unit. Wrote Leslie about the deluxe Pullman equipment,

> First, we are impressed with the smooth and delightful motion, and are told it is owing to a new invention, in the shape of paper wheels applied to this car, and incredible though the information sounds, meekly accept it, and proceed to explore the internal resources of our kingdom. We find everything closely resembling our late home, except that one end of the car is partitioned off and fitted up as a kitchen, storeroom, scullery—reminding one, in their compactness and variety, of the little Parisian *cuisines*, where every inch of space is utilized, and where such a modicum of wood and charcoal produces such marvelous results. (Quoted in Grant 1990, 33)

A wag had not been joking with Leslie about the Pullman car being equipped with "paper wheels," yet this reference aroused puzzlement and even smiles among contemporary travelers. This product involved the use of compressed paper or what was often referred to as "strawboard," being substituted for wood at the center of the car wheel. Even though greatly compressed, the paper center was spongy enough to cushion the ride and reduce the noise of the wheel moving over the rails. Introduced in 1869 by Richard Allen, a locomotive engineer turned inventor, paper wheels were widely used between 1880 and 1915, and were common on Pullman passenger equipment.

Although the Pullman Hotel Car did not have a long service span because of its high costs of operation, sleeping cars and dining cars became permanent features on major passenger trains during the post–Civil War period. Even though the Pullman Palace Car Company had competitors, the name became synonymous with the sleeping car. George M. Pullman, the firm's founder, did not "invent" the sleeper; he made it popular. Most notably, in 1856 Theodore T. Woodruff had received a U.S. patent for the upper and lower berth and that triumph proved to be one of Pullman's early business challenges. Pullman, who repeatedly traveled between Chicago and upstate New York, had encountered long, uncomfortable, and sleepless overnight journeys. Indeed, these experiences afforded him plenty of time to consider how nocturnal trips might be improved. After various experiments Pullman in 1864 introduced the *Pioneer*, the most luxurious car of its day and so big—a foot wider and 2.5 feet higher than any other passenger car in use—that there was hardly a railroad in the country that could operate it safely. Problems mostly lay with wooden station platforms, but they could be inexpensively cut back. The *Pioneer*'s twelve open sections, which could be easily converted into bedroom space, provided the greatest passenger comfort to that time. But if not for the assassination of President Abraham Lincoln in April 1865 and Mrs. Lincoln's desire to have the new car attached to the funeral train between Chicago and Springfield, Illinois, the $20,000 *Pioneer* might have become a classic white elephant. The positive exposure the *Pioneer* received, especially from the press, did much to bolster the fledgling Pullman company.

Following the Civil War, passengers who traveled only during the day began to have more opportunities to ride in something other than a utilitarian coach. Eastern roads, especially, had introduced the drawing-room car that then involved into the modern parlor car. These pieces of rolling stock offered comfortable, upholstered chairs that were liberally spaced, allowing for utmost comfort. Since an extra charge was applied for this space, "coarse and vile" males were unlikely to become riders, allowing a "safe" environment for female travelers and children. The undesirables customarily took seats in the "day coach."

ROLLING STOCK: THE MAIL CAR

While neither a freight car nor a passenger car, by the post–Civil War era train watchers might have noticed a specially designed piece of rolling stock that usually appeared near the head-end of a passenger train or occasionally in a solid train of like equipment. This was the U.S. Post Office or Railway Post Office (RPO) car. Early in the history of rail transport, governmental units in Britain, Europe, and the United States sent sacks of letters on regularly scheduled passenger movements. Beginning in 1838, the Grand Junction Railway of England started to sort mail en route; mail no longer was just merely being hauled.

In the United States the Hannibal & St. Joseph (H&StJ) Railroad played an important role in the development of mail by rail. This road, the

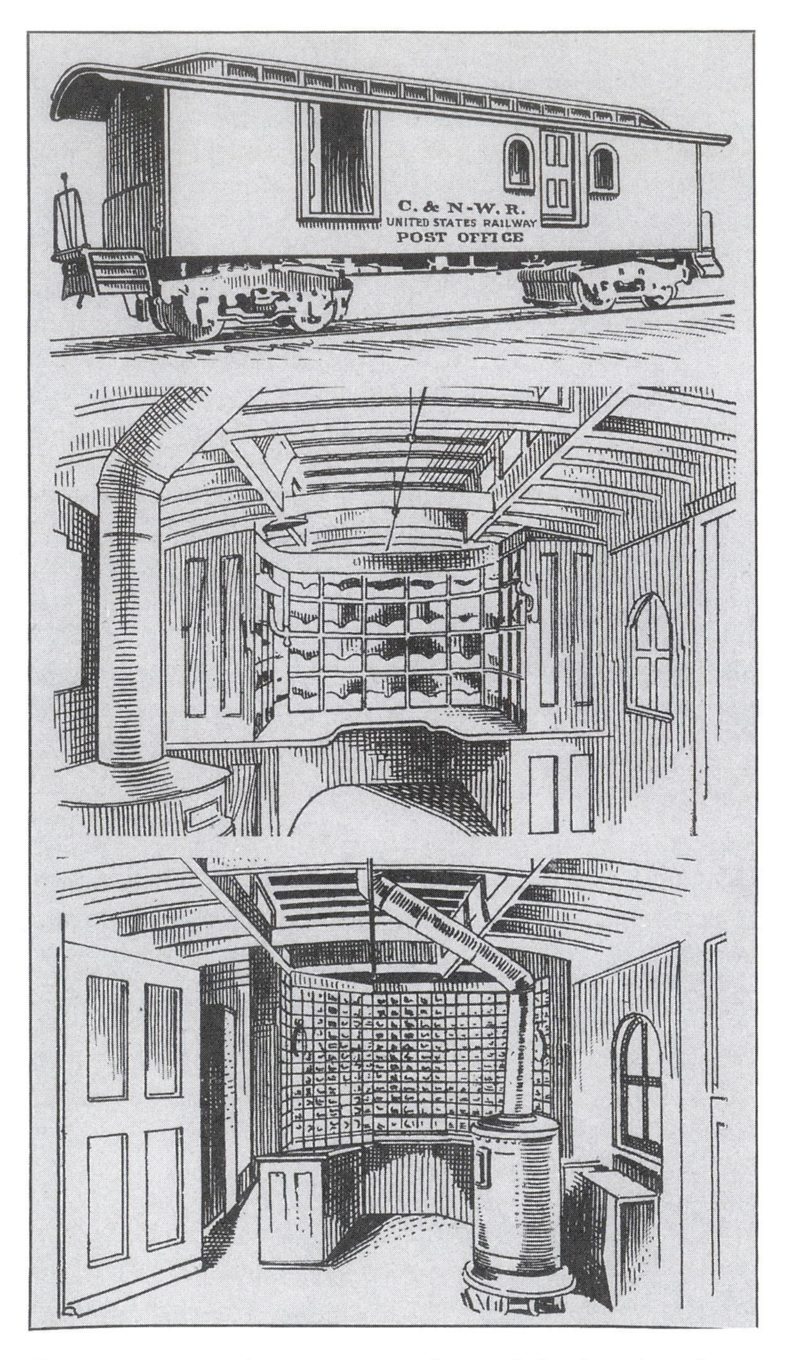

The inventive nature of Americans is vividly revealed in these three drawings of what was the nation's first Railway Post Office (RPO) car. The same type of compactness, which remained a hallmark of RPO equipment, also became common in sleeping and dining cars as well. Courtesy of the author.

first carrier to reach the Missouri River, quickly demonstrated its strategic importance. Prior to the building of the H&StJ, transcontinental mail had proceeded westward by steamboat and stagecoach. On April 2, 1860, however, a special train, hauled by a wood-burning locomotive, carried the mail from Hannibal to St. Joseph, Missouri, a distance of nearly 200 miles, in slightly more than four hours. Upon arrival, the oil-skin-wrapped letters were handed over to the rider scheduled to make the first relay on the inaugural westbound run of the Pony Express. Then, two years later, another epoch-making improvement was introduced when, on July 28, 1862, the first railroad car equipped for sorting the U.S. mail in transit ran westbound from Quincy, Illinois, to the Missouri River.

But it would be on the Chicago & North Western Railway where the Railway Post Office came of age. Troubled by the chaos of mail handling caused by the unprecedented volume of the Civil War, George B. Armstrong, an assistant postmaster in Chicago, conceived of the modern "post office on wheels." As he commented to a superior in 1864, "Passengers traveling over railroad routes generally reach a given point in advance of letters, when to that given point letters must pass, under the present system, through a distributing office, and when letters are subject to a distributing process in more than one distributing office as is largely the case now, the tardiness of a letter's progress toward its place of destination is proportionately increased." Concluded Armstrong, "But a general system of railway distribution obviates this difficulty. The work being done while the cars are in motion and transfers of mails from route to route and for local delivery on the way, as they are reached, letters attain the same celerity in transit as persons making direct connections" (Carry 1909, 14–15). During 1864 and 1865 the North Western built five "mail cars," and they operated north and west from the Windy City. And this company claimed operation of the first RPO cars in which mail was picked up, dropped off, and sorted all along its assigned route. It was appropriate that Armstrong became the first superintendent of the Railway Mail Service (RMS) when it was established in July 1869. By the 1880s hundreds of these special RMS cars, equipped with racks fitted with sacks and pouches, sorting cases, and tables, became common, even on some of the more remote lines. Some mail cars might be termed "palace cars"; they featured kitchen-dining compartments, washstands, beds or sofas, and state-of-the-art heating and ventilating systems and lighting. And by the 1880s specially hinged, cast-iron mail car furniture had been introduced.

The technology associated with the RMS involved more than specially constructed and equipped pieces of rolling stock. To facilitate mail movement RPO cars that were part of high-speed, limited stop runs or made up solid mail trains soon were picking up letters "on the fly." In 1867 an Ohio inventor received a patent for a crook-arm mail catcher. A station agent or post office employee simply tied a sack of mail on a trackside stand and a trained RPO clerk could use the arm to snatch the bag, even if the engineer had not reduced speed. For decades this device became a standard part of all RPO equipment, contributing to the success of the RMS and pleasing countless postal patrons.

COUPLERS AND BRAKES

As anyone who has ever thought about a freight or passenger train realizes, a fundamental dimension of railroad operations involves the makeup and breakup of a string of cars. Only the steam locomotive and tender are "slave" units that are semipermanently attached. When railroads turned their first wheels, the matter of a proper fastening device became paramount. In only a few cases did anyone want a *perpetually* attached set of cars. If that were to happen, the result would be negative, making switching impossible, necessitating different side-track configurations, and causing repair work to be more challenging and time-consuming.

Simplicity appeared to drive the technology of the earliest couplings in the United States. Indeed, in the mechanical world it is often best to embrace the most basic design, because of cost, ease of reproduction, convenience of handling, and the like. At first heavy metal chains were used, but they quickly gave way to primitive link-and-pin couplers that, in fact, remained the common car union on passenger rolling stock until the 1870s and on freight equipment until the 1890s. These primeval devices lasted into the twentieth century on some logging and mining roads. Although variations occurred, typical link-and-pin couplers seemed adequate since they worked reasonably well and became mostly standardized. Yet this coupling system proved to be laborious and dangerous for railroad workers. The brakeman was required to step between the cars as they were about to hit, lift the link that was already in the draft box (an iron device mounted in the center of the end sill of each car), and guide it into the draft box of the car to be coupled. He then had to drop the pin to hold the link and couple the cars. Not really knowing the amount of slack between cars, the brakeman and engineer found it difficult to predict the movement that occurred after the link was slipped into the draft box. This situation resulted in the hands or fingers of the trainmen frequently being smashed in the coupling process and, on occasion, serious injury and death.

Interestingly, railroads in Europe and Great Britain opted for coupling systems that differed dramatically from those developed in North America.

Over time these companies created a screw and turnbuckle arrangement for uniting rolling stock. The actual coupling devices were located at the center line of each car and employed buffers on the outside portions of the end sills. The coupling kept the rolling stock attached, while the buffers, which contained heavy-duty springs, prevented cars from crashing together when the train slowed or stopped. By the 1860s this coupling system was commonly employed in Europe while American railroaders dealt with the not-so-satisfactory link-and-pin couplers.

Fortunately, more adequate coupling devices developed in the United States, albeit somewhat slowly. At first inventors offered variations of the primitive link and pin. Their creations involved varying ways to guide the link into the draft box without having the brakeman do it manually, usually employing some type of metal buffer mounted on the end-sill near the draft box in order to reduce annoying and dangerous slack. Other largely experimental couplers used a metal, fixed tongue mounted in the draft box and featured a hook that engaged the tongue from the mating car.

The dramatic breakthrough in the technology of car coupling occurred with the development of the "Janney" or "knuckle" coupler. Following the Civil War, Eli Hamilton Janney, a mechanically talented former Confederate Army officer, fashioned a coupler that resembled the knuckles of the human hand. After successful testing by a subsidiary of the Pennsylvania Railroad, the Janney Car-Coupling Company by the 1880s sold what would become the standard automatic coupling device. The firm offered a pivoting forged iron or cast steel knuckle that permitted two couplers to slide past each other. The force of the contact between two cars actuated a spring that closed the knuckles, thereby joining the cars. Later, for use on freight equipment, a heavy drop pin, which gravity held in place, replaced the less satisfactory spring. Moreover, the strength of the knuckles themselves increased. And significantly, the installation of a metal bar, which extended to the side of the car, allowed the knuckles to be released easily. This feature eliminated the need for the trainman to step between the cars and predictably resulted in fewer injuries and deaths. Another attractive feature of the coupler was that such an equipped piece of rolling stock could accommodate a car that had the old-fashioned link-and-pin mechanism. At last, American railroads had access to a durable and largely trouble-free coupler. Somewhat surprisingly, other national railroad systems did not commonly use the Janney invention, largely being satisfied with the functional buffer-screw-turnbuckle mechanisms.

Hand-in-hand with the development of better car couplers came dramatic improvements with braking devices. During the gestation phase, stopping either a freight or passenger train was not a major problem. Pioneer steam locomotives pulled short trains and operated at slow speeds. An experienced engineer commonly allowed his train to drift to a halt, or he slowed or stopped the train by throwing the locomotive into reverse. If required, the engineer and crew members could push a lever-type hand brake against the wheels of the locomotive, tender, or cars. In the 1840s an important advancement in braking trains came about with an improved linkage mechanism that allowed the engineer to reverse his locomotive while operating at speed. About the same time, it became possible for a brakeman to set the brakes on all the wheels of a car with a single brake wheel. As the trainmen turned this device, chains or levers pressed the "brake shoes" (originally blocks of wood and later iron) against the wheels. A combination of reversing the engine and setting brakes on the rolling stock allowed for better control over trains that grew longer and ran faster.

Yet the "double-acting" brake arrangement was hardly perfect. The repeated reversing of the locomotive produced enormous wear and tear on the linkage and driving wheels. Moreover, this braking damaged rails, particularly at station stops. And for brakemen, who had to scamper along swaying freight cars, the job was both difficult and dangerous. Recalled a trainmen who worked on the St. Louis–San Francisco (Frisco) Railway, braking was not a pleasant chore, especially in bad weather:

> By the time we [i.e., the train] got to Stirling [Missouri] and started down the hill to Willow Springs there was so much ice on the rails that the engine could not hold them [cars], for when the hoghead [engineer] would hoss her over into the back motion, the crushed ice on the rails would cause the drive-wheels to lose their grip on the rails and spin around backwards, and as . . . [the locomotive] could not hold anything that way, the six cars and the caboose shoved her right along, so the hogger [engineer] grabbed the whistle-cord, and began calling for brakes, and I took my lantern and scrambled back over the tank [locomotive tender]. And I had one heck of a time getting from the back end of the tank over onto the first box car, for everything was covered with ice, and when I finally did get upon top of the box car I could not stand up and walk, for by this time we were going so fast and the car was rocking from side to side so bad that I had to get down on my hands and knees and crawl to the other end of the car where the hand brake was at. After setting it up as tight as I could, I managed somehow to get over onto the next car, but the ratchet dog on that brake would not hold, so I did not try to go any further, but just sit there on the running board, with my lantern between my legs, and held on for dear life. (Grant 1991, 103–104)

Fortunately for the Frisco, this particular brakeman was a conscientious employee. Some of colleagues, however, were not. Indeed, this arduous and risky work commonly attracted a poor class of men, the most negligent of whom caused endless trouble. A repeated criticism was leveled at the lazy brakeman who, rather than applying all the brakes in his charge, bore down hard on the one or two closest at hand. This response could break the brake-beams, causing a derailment or locking and thus damaging the wheels. In time, industry representatives came to see that a better brake technology, which would eliminate the need for trainmen moving from car to car, would be money well spent.

By the 1870s mounting wrecks, deaths and injuries, property damage, and other negative factors caused by deficient braking devices led to a spate of applications to the U.S. Patent Office for train brakes. In fact, this federal agency had issued more than 300 patents. The vast majority involved mechanical methods that would provide a *continuous* train-braking system and included such approaches as compressed air, hydraulics, spring, steam, and vacuum.

Although George Westinghouse, who possessed a keen innovative mechanical mind, did not "invent" the modern train brake, he did much to perfect and market it. He wisely took the air or pneumatic approach. In 1869 he tested on the Pennsylvania Railroad his first version, popularly called the "straight air brake," and his equipment worked reasonably well. Braking occurred when air, generated by a steam-driven storage tank placed on the locomotive, rushed through a series of hoses and pipes that ran the length of the train. When activated by the engineer, this air triggered a brake cylinder under each car where a piston moved levers and rods that pressed brake pads against the moving wheels.

But early on Westinghouse, mechanical officers, and operating personnel realized weaknesses. If a major leak occurred in the lines that carried the compressed air, there might be inadequate air pressure to activate the brakes. Also, if the train broke apart, those pieces of rolling stock not attached to the locomotive became brakeless. Then in 1873 Westinghouse devised a method whereby air pressure was used to keep the brakes *off* rather than to put them on. The basic part of the altered system was the *triple valve* that was designed to be applied to the locomotive and every car. The valve was aptly named, for this device carried out three distinct functions: "[It was] a pressure-sensitive valve that could admit air from the train line into the reserve tank, readmit air from the same tank to the brake cylinder, and exhaust the air to release the brake." To make the Westinghouse system work, each car needed such a valve as well as its own air tank. "It was a remarkable chunk of cast iron no larger than a grapefruit," noted historian John H. White Jr., "and so amazing to mechanics of its day that some of them half expected it to speak" (White 1978, 552).

The earliest triple-valve air brakes worked well for passenger rolling stock, but they were not always effective on long freight trains. Consequently, during the summers of 1886 and 1887 exhaustive test runs were conducted by the Master Car Builders Association, an industry professional organization, on the "Burlington Hill," located on the main line of Chicago, Burlington & Quincy Railroad near Burlington, Iowa. As a result of these trials, Westinghouse and his associates developed a faster, heavy-duty triple valve that met all of their requirements. The key was the addition of vent pipes on each car that increased the speed of the system sixfold. These changes meant that brakes on each car activated almost simultaneously.

Still, contemporaries of Westinghouse toyed with other braking systems. The most significant was the vacuum brake, which a Watertown, New York, inventor, Frederick W. Eames, perfected. It was a straight braking system; specifically, it employed a steam ejector to develop the vacuum required in the train lines to reduce speed or to stop. "Braking was created by a pulling action opposed to the pushing action of the Westinghouse automatic air brake," observed historian Robert L. Frey. "All cars and locomotives had to be equipped with vacuum 'pots,' an addition which restricted the interchange of cars once most railroads adopted the Westinghouse brake" (Frey 1988, 31–32). Yet the Northern Pacific Railroad, for one, used the Eames vacuum brakes on its most mountainous lines. And significantly, the vacuum brake system became popular outside North America.

Even though major technological breakthroughs had occurred with Janney couplers and Westinghouse air brakes, the American railroad industry was slow to adopt these improved devices, especially the air brake. Cost was the principal roadblock. Although the price tag for fitting passenger cars with automatic couplers and air brakes was reasonable due to the relatively small number of pieces of equipment involved, the freight car situation was dramatically different. A major carrier did not fuss much about spending roughly $100 each for installing air brakes on a few hundred passenger cars, but doing the same to 40,000 or 50,000 freight cars was another matter. The charges would have been less per unit had it not been for the patent monopoly held by Westinghouse. Railroad management also realized that even with these heavy expenditures, air brakes did not reduce the work force required to operate trains; after all, trainmen still needed to check equipment, couple cars, throw switches, "spot" equipment, and the like. This technology was hardly labor saving.

The catalyst for adopting these two pieces of replacement technology came amazingly from the crusading efforts of a liberal Iowa minister, Lorenzo S. Coffin. As an appointed member of the Iowa Railroad Commission, this public-spirited official, by riding frequently on both freight and passenger trains throughout the Hawkeye state, had learned about railroad employees who had lost fingers, hands, and even their lives trying to make couplings with link and pins. Furthermore, he had become aware of the dangers of attempting to stop trains with hand brakes. He demanded that carriers adopt the knuckle coupler and air brake immediately. As Coffin rhetorically asked, "Why not have self-couplers, that will couple automatically when the cars are pushed together, and not require the men to go between the cars; and why not have air-brakes on freight as well as on passenger cars, under the control the engineer, as on the passenger trains?" (Coffin 1903, 562). Just as he had battled for abolition and prohibition, he focused his attention on the newly created Interstate Commerce Commission (ICC), and then Congress and the White House.

Finally, Coffin won. On March 2, 1893, President Benjamin Harrison signed the Safety Appliance Act. This federal statute gave railroads five years to equip all of their rolling stock with couplers that would automatically lock upon impact and to install air brakes. But the next five difficult years that resulted from the Panic of May 1893 ruined railroad revenues, throwing many carriers into bankruptcy. As a result, compliance faltered. The industry won an extension and August 1, 1900, became the new deadline. Fortunately for public and employee security, most rolling stock finally featured these safety devices. Railroad management eventually came to realize that accidents were reduced and the wear and tear on equipment were less. The Safety Appliance Act turned out to be a classic "win-win" position. As Coffin himself observed, this legislation had a powerful impact. "In the year after President Harrison's approval of the measure there were 2,837 railroad men killed and between 20,000 and 30,000 injured. Four years later there were only 1,693 killed, making a saving of 1,044 lives, and there were about 5,000 fewer men injured by accidents" (Coffin 1903, 580).

SIGNALS

Just as early inventors and others interested in things mechanical did not concentrate on couplers and brakes, they initially ignored matters of signaling. At first, demand simply did not exist. Daily trains movements were relatively few and generally slow. A passenger or freight train regularly departed from a terminal at a particular time, and crews knew usually when and where "to meet" any other train on the single-track line, entering one of variously installed passing sidings. A locomotive engineer, however, continually used his eyes and ears, watching for a plume of smoke or at night for a flickering headlight and always listening for the steam whistle, to learn the approximate location of an oncoming train. What the railroads then employed is best described as the "time-interval" method of train control. Management always drilled into the minds of trainmen that they should be cautious, especially if a delayed oncoming train had not arrived at the prescribed meeting place.

Pioneer efforts to prevent accidents also involved the use of an individual's hands and forearms and the employment of flags. When the first train traveled over the Stockton and Darlington Railway in the mid-1820s, for example, a "signalman" rode on a horse ahead of the locomotive and carried a large flag. Yet, such a labor-intensive response to the matter of train control did not take root in most places, including North America. The traveling signalman really was being too cautious.

Yet, as railroads evolved, "fixed" signals became useful, even mandatory. By midcentury, the network of iron rails had grown to such an extent in Europe and North America as to make such devices highly desirable. Perhaps the most representative was the ball signal that was commonly placed where rail lines crossed one another. Usually known as the "high-ball signal," this rather prosaic instrument consisted of a canvas ball hauled to the top of a wooden pole by a rope. When hoisted by a member of a train crew, a station agent, or another railroad worker, the ball indicated "all clear," and when lowered, it meant "danger." At night an ordinary oil-burning barn lantern replaced the ball. Although admittedly primitive, high-ball signals continued to be used worldwide until the 1930s and in a few places even later, attesting to their simple practicality.

Surely, the most important technological advancement of the Youth Period involved the magnetic electric telegraph. Even though commercial telegraphy began in the mid-1840s, it took considerably longer before it became part of train-control practices. The Erie Railroad, then the New-York & Erie, which throughout its long history revealed an innovative bent, gained much positive notoriety when in September 1851 its resourceful general superintendent, Charles Minot, conceived the idea of using the recently perfected telegraph for traffic control. Soon thereafter Minot placed a dispatcher with access to telegraphic communications in charge of train movements for each operating division, something that within a score of years became common railroad practice. Perhaps the big breakthrough in the employment of this technology came during the Civil War when the U.S. Military Railroad extensively used telegraphic train dispatching.

What Minot demonstrated was that a dispatcher could not only contact train crews via trackside operators, but could also keep in communication with towermen and anyone else who could assist in keeping trains a safe distance from one another. As part of this telegraphic network, railroads quickly erected another type of fixed signal, namely, signal masts at depots and train-control towers. When a train approached, engine crewmen could determine if they needed to stop for a message, generally referred to as a train order. Or they might learn of the need to pause for freight and passengers. By the position of the signal or by an oil-burning lamp at night (red meant to stop and white to proceed), they knew how to respond.

Although the signaling system that evolved after Minot's brainstorm worked well for lightly and modestly active lines, it was not wholly suited for busy main lines, especially those in congested urban centers. Admittedly, by the 1880s most rail lines had only one or two passenger trains and perhaps as many freight trains, although seasonal variations occurred, especially for the movement of foodstuffs. Those interested in improving safety and the general flow of traffic and in saving labor costs wanted some device or system that would not be dependent on dispatchers and trackside operators. After all, these individuals might fall asleep, become inattentive, or face a distraction; a fail-safe mechanical system, devoid of human control, appeared to be the logical alternative.

Following the Civil War, the quest for a dependable "automatic" signaling system became ongoing. Essentially what inventors sought to do was to create an automated block system of signal controls. Specifically, when a train entered a particular stretch of track, it would trip a signal to red, warning trains in the opposite direction to stop. And when that train left the block, the signal would turn to white (later, green became the color for "all clear"), indicating that it was safe for the train to continue. A variety of schemes appeared in this "space interval" approach, including one developed by Thomas Hall, a Connecticut tinkerer. He opted for wheel-actuated treadles that were placed next to the railhead to trigger an electrical switch that prompted a solenoid to control the positions of trackside signals.

Yet, railroads were slow in the installation of any form of block signaling. The reasons were hardly surprising: costs were considerable, labor unions did not make demands, and the public was clearly more interested in automatic brakes. If the latest signal technology was adopted, it usually involved trackage near congested terminals. The twentieth century, however, became the time of the automatic signaling. Some railroads, because of heavy traffic considerations, decided to double track their busiest lines, and this allowed the manual systems of train control to work reasonably well. After all, a single-track operation was much less forgiving than a double-track configuration.

Part of the development of trackside signaling included the creation of what the public also considered to be highly important, namely, automatic public-road-crossing warning devices. As railroads developed, passive signs appeared to warn travelers of tracks and trains. The familiar crossbuck sign dates from the nineteenth century, the center of the X being 9 feet above the ground so as to be at eye level of a person on horseback. And railroads, especially in urban areas, hired individuals, often crippled or elderly employees, to flag street crossings and to physically lower and raise crossing gates. But flagging and hand-mechanical operations were expensive and never foolproof. But with the increased use of electricity in railroad operations, it is not surprising that inventors exploited this rapidly developing technology. In the mid-1880s two Barretts Station, Missouri, telegraph operators received a patent for an automatic bell warning device. The mechanics of this "Electric Railway-Signal" were wonderfully simple: an electrical track circuit, when completed by an approaching train, automatically operated a warning bell. This sounder would work from the moment a train reached the circuit from one direction and continued activation until the last car cleared the circuit in the opposite direction from the crossing. This patent also had a feature by which the warning device ceased operation as soon as the rear car cleared the crossing. Yet it reactivated if another train approached from the same direction as the first train, but while the preceding train was still in the circuit. The concept remains in use for both bells and crossing gates.

TIME

During the Youth Period, matters of time became vital to railroad operations. When safety, efficiency, and convenience were considered, it became mandatory for railroad employees (and the public, too) to know the precise time of day. Fortunately, during the mid-nineteenth century and later the quality of time pieces, whether station clocks or pocket watches, improved and their overall costs declined. As train traffic and speeds increased, management became committed to ensuring accurate timekeeping. "Each Engineer will be furnished with a watch which shall be regulated by the Station Agent at the commencement of each trip," was a statement that appeared in the rule of operations book published by the Pennsylvania Railroad in 1849. At times it was not uncommon for a railroad to supply pocket watches and to hire jewelers and other experts to check these instruments regularly for accuracy. If timepieces were not issued—a practice that mostly died out after the Civil War—trainmen and other specified personnel were required to acquire a quality pocket watch, usually with at least seventeen "jewels," small gems or gem-like bits used as bearings, that would then be inspected by a railroad-approved jeweler for reliability.

Establishing the correct time became paramount. To properly adjust their station and terminal clocks, railroads took advantage of the noontime signal supplied from the U.S. Naval Observatory, located in Washington, D.C., or perhaps by private observatories that sold regular time signals. It was customary for a company to transmit its operating time once daily from a central location, and for station agents and other office employees then to adjust their local timepieces accordingly. By the early 1880s, the development of synchronizer-equipped clocks allowed a road's chief dispatcher to be reset automatically (usually on the hour) to the correct time. By the end of the century the Western Union Telegraph Company became the principal disseminator of the precise time, acquiring this information directly from the Naval Observatory.

The time problem, however, had not been fully solved with quality clocks and watches and the employment of accurate time signals. What existed was a crazy-quilt pattern of dozens of times based on the position of the noonday sun. For instance, when it was noon in Chicago, it was 11:27 AM in Omaha, 12:31 PM in Pittsburgh, 11:50 AM in St. Louis, and 12:17 PM in Toledo. Before standard time zones existed, a railroad handled its train movements by using the local time of the major city that it served. Indeed, by the mid-nineteenth century there were about 100 *different* city times used by the railroad industry. Yet many communities along a particular carrier would have other times, for they employed the system of local time and that was set by location. Initially, this was not much of a concern for a railroad and its patrons, but as companies expanded through merger, lease, and line construction, the time issue drew increased attention. After the Civil War travelers who found themselves in Buffalo, New York, for example, would discover that local railroads used *three* different times: Albany time for trains operating on the New York Central & Hudson River, Columbus (Ohio) time for service on the Lake Shore & Michigan Southern, and Buffalo time for the remaining carriers. "The traveler's watch was to him but a delusion," observed an educator from Saratoga Springs, New York, and an early advocate for a standard time. "Clocks at stations staring each other in the face defiant of harmony either with one another or with surrounding local time and all wildly at variance with the traveler's watch, baffled all intelligent interpretation" (quoted in Bartky 1983, 14). Even operating personnel could become confused: a trainmen might accidentally look at the wrong clock and adjust his watch incorrectly.

It would be the railroad industry rather than the federal government that launched the system of time zones. After a long series of discussions, William F. Allen, managing editor of the *Official Guide of the Railways*, successfully argued his plan for a four-zone system before what was called the General Time Convention. The main features of the Allen scheme involved specific rail line assignments, time changes "at points where they change at present, and at the terminus of a road, or at least at the end of a division," and consistent with the general tenor among the astronomical observatories that sold time to the railroads—an expectation that "local time [of cities] would be practically abolished" (quoted in Bartky 1983, 18). The group endorsed the proposal, and on Sunday, November 18, 1883, almost 600 railroads dropped the fifty-three arbitrary times they had been using, and adopted Greenwich-indexed meridians that defined the times in each of four new zones: Eastern, Central, Mountain, and Pacific. Generally speaking, these zones ran along the 75th, 90th, 105th, and 120th meridians. Carriers in Canada, too, embraced the new times zones, although the Maritime Provinces on Canada's east coast adopted a fifth zone, known as Intercolonial Time. News of the new standard time arrangement was greeted with some derision, but rather passively and quickly accepted by the public. In the following year, 1884, an international conference established twenty-four time zones worldwide. Then in 1918 the U.S. Congress intervened and adopted a set of standard time zones that closely paralleled what the railroads had already established. At that time, too, the concept of daylight savings time was first introduced, largely out of wartime necessity.

Whether a railroad company used sun time or standard time, it needed to convey travel information in an appropriate way. In the earliest years, carriers depended on broadsides, posters, newspaper notices, and even word of mouth to inform the public about train schedules. About the time of the Civil War, railroads changed to a table form, a "time table" that traced a specific passenger train and its times of arrival and/or departure station by station. As the century progressed, timetables, maps, and promotional brochures, especially for major roads, became more elaborate, and many included fancy illustrations. As a revolution in printing technology occurred, unit costs plummeted. Wood pulp paper, wax-engraving techniques, inexpensive lithographic processes, and high-speed printing presses collectively had a positive impact on railroad publications of all sorts.

The process of technological improvements seemed unending. Although the federal government got involved with regulating railroad safety and later matters of time, it was the railroad enterprise and those industries that supplied materials and products that revolutionized what by the end of the nineteenth century had become the dominant form of land transportation. In 1889 Charles Elliot Perkins, president of the CB&Q Railroad, expressed this commitment to improved technologies: "If one thing

I am very certain, that is that the character of the service is becoming of more and more importance, and we must keep up with the march of improvement in this respect or we shall lose traffic" (quoted in Cochran 1953, 141). Betterments had hardly ended; merely Youth was about to turn into Maturity.

3

Maturity, 1880–1940

STEAM LOCOMOTIVES

As the twentieth century dawned, train watchers worldwide noticed that steam locomotives were becoming larger and often were traveling much faster than they had a decade or two earlier. In the United States, notably, except on branch lines and shortlines, little steamers had overwhelmingly given way to the powerful giants of the maturing Railway Age. Along with the inborn desire on the part of railroad personnel and the principal manufacturers to perfect bigger and better locomotives was the fact that the public, especially in the United States, wanted the power and speed that such engines could provide. "Our railroads kill their thousands every month in wreck or trespass," editorialized a popular, mass-circulation magazine, *World's Work*, in March 1907. "In more than half the cases, the real truth underlying the tragedy is the fact that the train was running at forty or fifty or sixty miles an hour over tracks that were built for trains that never ran but thirty miles an hour. The people demand it. The railroads must obey. Each year, the manufacturers of locomotives are called upon by the big lines to produce and deliver more and more engines that can haul a ten-car passenger train at sixty miles an hour. Even in the far South or in steady old New England, the cry is ever for more speed." Concluded the writer, "The railroads take big risks. They have to. Competition grows terrible, and the railroad, like the individual, must live" (*World's Work* 8, 595). This essay, "Pace That Kills," appropriately addressed the issue of speed, and no one challenged this perspective.

In what business historian Albro Martin has appropriately labeled the "second building" of American railroads, following the cataclysmic depression of 1893–1897, companies spent heavily to improve their physical plants to handle a substantial increase in passenger volume and freight tonnage. This was the time that roads, often through jointly owned terminal companies, erected truly magnificent monuments to the Railway Age, ranging from Pennsylvania Station in New York City to Union Station in Kansas City, Missouri. Both facilities were part of urban betterment movements that emphasized the importance of beauty and efficiency. Less obvious to most citizens were line relocations, larger switching yards, public-road grade separations, replacement signals, and other improvements.

These betterments did not mean that the process of building lines had come to an end. In 1898, after nearly seven decades of construction, the national railway network totaled 245,300 miles; ten years later, track mileage reached almost 328,000, a 34 percent gain on top of a fully mature base. Route mileage would then peak on the eve of World War I. Similarly, the number of pieces of rolling stock substantially increased. Take locomotives. In 1897 the national fleet totaled approximately 36,000 units, and a decade later there were nearly 55,000 in service.

When railroad companies sought to communicate to citizens their commitment to a better transportation network, the roads inevitably extolled their

In the age of specialization, steam locomotive manufacturers designed motive power that met special needs, including this 1903 Brooks-built high-speed Camelback-type 4-6-0 for commuter service on the Long Island Railroad. The Camelback design permitted the fireman to shovel anthracite coal directly into the extra-wide firebox and gave the engineer, whose cab was toward the front, better visibility and fewer distractions. Courtesy of the author.

latest motive power. The largest carriers, most of all, repeated these messages of modernity in newspaper and magazines advertisements, public timetables, name-train folders and travel brochures, and lavish displays at fairs and special events. The Pennsylvania Railroad (PRR), for example, which by this time immodestly presented itself as the "Standard Railroad of the World," spent heavily on a new locomotive and a state-of-the-art locomotive testing plant as the centerpiece of its well-attended exhibit at the Louisiana Purchase Exposition in St. Louis, Missouri, in 1904. Visitors to the Hall of Transportation seemed awed by the size of the locomotive and engrossed when a team of technicians carefully measured fuel consumption, pulling power, and tractive resistance. Everyone surely understood that this was the "Gear and Girder Age," a befitting description coined in the 1980s by historian Cecelia Tichi.

The increase in the size and hence power of steam locomotives was impressive. In 1890 the largest product produced by the Baldwin Locomotive Works, the nation's premier engine manufacturer, weighed approximately 185,000 pounds in running order; sixteen years later the Great Northern Railway acquired five Mallet compound engines (2-6-6-2) that individually topped the scales at more than 350,000 pounds. Because locomotive power capacity was proportional to weight, these behemoths on the Great Northern and on other carriers did much to improve productivity by pulling heavy payloads, especially over rugged mountain grades.

To trackside observers, the Mallet seemingly represented the logical fruits of decades of designing and building ever bigger and more pieces of motive power. In the pre–World War I era several American carriers, including the Baltimore & Ohio; Carolina, Clinchfield & Ohio; Chicago Great Western (CGW); Erie; and Great Northern, tried Mallets. These "Snakes," as railroaders sometimes called them, were named after Swiss-born inventor, Anatole Mallet, who in the latter part of the nineteenth century built the first articulated engine and received a patent for a compound locomotive. Essentially what Mallet had done was to place two steam engines under a single boiler. The rear engine was a high-pressure type that was attached rigidly to the frame, while the front low-pressure engine pivoted so that on sharp curves it would swing independently of the rest of the locomotive. Different from "simple" steam engines, the Mallet compound had its high-pressure cylinders power the rear set of drivers, and the steam exhausted into the much larger lower pressure cylinders located in the pivoting front engine. (The increased size of the low-pressure cylinders was required to compensate for the lower average steam pressure.) These locomotives, which came with various wheel arrangements—0-6-6-0, 0-8-8-0, 2-6-6-2, and 2-8-8-0—were usually used to "drag" long trains over steep trackage and also to perform heavy switching assignments.

The Mallet concept produced an enormously powerful locomotive. These monsters could easily pull a 4,000-ton train, while a typical contemporary freight engine usually could handle only 1,500 to 2,000 tons. As might be expected, there was a price to pay. Mallets were exceedingly slow, prone to derail, hard on the track, and expensive to maintain. These limitations explain why some companies, for example the CGW and Erie, sold off their Mallets rather quickly. Moreover, by 1915 the popular use of superheated steam greatly diminished the need to extract additional power from the low-pressure region of expansion.

Indeed, "superheaters" did much to "soup up" conventional steam locomotives. In the late 1890s a mechanical engineer equipped two locomotives of the Prussian State Railways with these devices. The company observed an approximately 25 percent increase in efficiency, and so it applied the device to other engines. The Canadian Pacific Railway (CP) noted these promising experiments and at the turn of the twentieth century sent one of its locomotive experts to observe these improvements, resulting in the equipping of a ten-wheeler with a smoke box superheater. Within a few years, other railroads in North America followed suit.

What was being perfected was a simple two-chambered manifold that contained no moving parts. Hairpin-shaped metal tubes, located at the top of the locomotive smoke box, ran into the flues, reaching almost as far as the rear tube sheet. Saturated steam traveled through the tubes, absorbing more heat from the exhaust gases, and then returned to the superheated side of the manifold, from which it was piped to the cylinders.

Although not many railroaders encountered the mighty Mallets, after the turn of the twentieth century more employees worked not only with locomotives that sported superheaters but also with ones that burned oil rather than coal. This replacement power source occurred for several reasons: growing availability of inexpensive oil, especially following the post-1901 petroleum boom in the American Southwest; increased convenience of this fuel source, particularly in transport and storage; and reduced workload for the firemen who, before the advent of mechanical stokers, shoved literally tons of coal during their daily runs. A fireman for the Atchison, Topeka & Santa Fe recalled his first contact with oil-burning locomotives in 1909: "They [Santa Fe] had some of the biggest engines there [Needles, California] that I had ever seen up to that time, but they were oil burners and all a fireman had to do was just to set up there on the seat box and watch his water and steam gauges, and operate the firing valve, which was a lot better than standing down on the deck heaving in the black diamonds" (Grant 1991, 171). By the time this veteran fireman worked his first oil-burning locomotive on the Santa Fe, other railroads continued to use mostly soft coal and some, almost exclusively in the Mid-Atlantic states, burned anthracite. These carriers collectively had invested heavily in coal technology and the required infrastructures, and they also usually maintained close connections with the coal-mining industry. Yet the petroleum age had dawned and more oil-burning locomotives made their appearance, much to the pleasure of both engine crews and everyone who detested thick plumes of black coal smoke.

In the twentieth century those firemen who worked on conventional coal-burning locomotives gladly welcomed the growing use of mechanical stokers. As engines became larger, it became virtually impossible for firemen to provide enough fuel to keep up the steam pressure; therefore, the development of a practical way of feeding the firebox became paramount. In time stokers, employing a large screw-tread feeder, sent coal into a crusher that then sprayed this fuel onto the fire. A similar technology was employed in home and commercial furnaces.

With all technological advancements, there were exceptions. This became the case with fuel. Some shortlines in the South remained wedded to wood; it was readily available and dirt cheap. Furthermore, labor costs were low. As late as the World War II era the Mississippi & Alabama Railroad, for one, used pine knots from along its 17-mile line for its several small steam locomotives and maintained "wooding-up platforms" at 5-mile intervals. "The wood burner, like home-baked bread, the ice box and the straight-edge razor," commented a visitor to 1947, "is still a symbol of self-reliant individualism in a time when daily and on every hand progress and industrialism demonstrate their multiple insufficiencies" (Beebe 1947, 19). Probably, though, financial considerations and not a brand of rugged Southern individualism explain this shortline's fuel choice.

While the little wood-burning engines of the Mississippi & Alabama chugged through the piney woods, locomotive designers were busily creating what experts considered to be the best "modern steam locomotive." Although historians of steam traction debate the precise moment when such a piece of motive power appeared, the time was likely the early 1920s and the place was the Lima Locomotive Works in the western Ohio factory city of Lima. (Earlier this firm had specialized in a geared logging locomotive known as the "Shay," named for its inventor, Michigan lumberman Ephraim Shay.) The skilled technical personnel at this relatively small yet innovative company believed that steam power capacity did *not* ultimately depend on size. Not too long before, engineers at the mighty Baldwin Locomotive Company took a different view, creating a 2-8-8-8-4 (but featuring only a modest 108 square feet of grate area) for the coal-hauling Virginian Railway and toying with the concept for a "Quintuplex"

By the 1930s railroad officials commonly wished to show off their industry's technological advances, especially with motive power. In preparation for the Century of Progress Exposition in Chicago, the Chicago & North Western Railway proudly displayed the evolution of steam locomotives from the 10-ton 4-2-0 *Pioneer* (right) to the mighty 409-ton 4-8-4 *Class H* locomotive (left). Courtesy of the author.

locomotive—a two-cab 2-8-8-8-8-8-2. This firm's philosophy for a better freight locomotive involved additional driving axles and more adhesive weight. Lima, on the other hand, believed that adequate boiler capacity and especially ample firebox size would allow sustained horsepower production at competitive freight-train speeds.

The prototype locomotive of Lima engineering became an ultramodern Mikado-type 2-8-2. In 1922 this engine, No. 8000, joined the motive power fleet of the New York Central (NYC) and successfully powered its deluxe passenger trains. Within a few years Lima developed what it called the A-1 type locomotive that originated the 2-8-4 wheel arrangement so as to place design emphasis squarely (and forever as it so happened) on horsepower and steam supply. "No other steam locomotive built after World War I had such a remarkable, lasting, and beneficial effect upon the American scene," noted a railroad journalist. "Simultaneous experimentation by other builders and roads on high-pressure boilers, compounding, and 3-cylinder transmission all came to nought" (Morgan 1959, 5). The 2-8-4s performed well in both passenger and high-speed freight service and established Lima as a major player in locomotive research and production.

The triumphs at Lima did not mean that other major improvements in steam locomotives had largely ended. In the late 1930s, for example, the PRR, working with the major commercial builders, created a four-cylinder, nonarticulated, 84-inch-drivered 6-4-4-6 locomotive with steam cylinders both front and center. The pilot and trailing trucks had six wheels each, thus giving the locomotive a novel wheel arrangement. At the World Fair's, held in New York City in 1939–1940, the railroad told visitors that this monster was "the world's largest, fastest, most powerful passenger steam locomotive." And the PRR was correct.

No matter the locomotive design, another major betterment firmly took root, namely, rolling bearings. In the early 1930s, the Canton, Ohio–based Timken Roller Bearing Company, working closely with speciality suppliers, bought a 4-8-4 locomotive from the American Locomotive Company and replaced traditional friction bearings with high-alloy steel bearings. In the two-year demonstration that followed, this Timken engine, known as TRBX 1111 (or widely called the *Four Aces*), traveled nearly 120,000 miles without any kind of roller-bearing troubles on any of its axles. Industry leaders opened their eyes to this technological accomplishment. The public, too, was highly impressed with a clever public relations stunt: three female PRR office employees in Chicago, working as a team, pulled this monster 700,000-pound machine forward a few feet. This female feat was possible, explained a Timken spokesman, because the rolling resistance of a roller-bearing-equipped engine was an astonishing 10 ounces per ton.

Even though the Timken triumph was never questioned, the addition of roller bearings on American steam locomotives was neither immediate

As the diesel age dawned for passenger-train service, several major railroads decided to upgrade their steam passenger motive power. In the late 1930s the Chicago & North Western Railway turned to the American Locomotive Company to construct a powerful 4-6-4 locomotive that featured metal shrouding that produced a highly "streamlined" appearance. Courtesy of the author.

nor universal. The negative financial impact of the Great Depression served as a severe roadblock. By the latter part of the 1930s, however, most new engines, designed for high-speed, long-distance service, sported the Timken product. TRBX-1111 did not remain in Timken hands; the company sold the locomotive to the Northern Pacific, where as No. 2626 it operated well into the diesel-electric era.

When railroads ballyhooed their latest steam locomotives, they would inevitably broadcast the adoption of roller bearings. In the late 1930s, for example, the Chicago, Burlington & Quincy Railroad (CB&Q) announced introduction of *The Aeolus*, a 4-6-4-type locomotive that it claimed to be the "World's First Stainless Steel Streamline Steam Locomotive." In advertising copy the railroad noted that "a major obstacle in operating steam locomotives at very high speeds has always been the terrific forces set up by the up-and-down and back-and forth motion of the driving rods and attendant reciprocating parts." Company engineers and technicians, working at the West Burlington, Iowa, shops, successfully responded to these challenges. Steel roller bearings did much to allow the smart-looking *Aeolus* to reach high, sustained speeds, becoming "practically vibration less at 100 miles per hour, and one that pounds the rails only a third as hard as conventional equipment." Without doubt the roller bearing had much to offer modern steam motive power (*Aeolus* 1940, 1–2).

Whether a steam engine represented old or new technologies, railroads might rely on in-house chemists and other experts to deal with the age-old problem of water quality. High mineral content often reduced efficiency and at times greatly impaired operations. For railroads that operated in the trans–Missouri West especially, water supplies often contained significant amounts of calcium carbonate and magnesium sulfate. Calcium carbonate was the major irritant because it caused boiler water to foam, exaggerating the level of water and sometimes allowing water to bubble over into the cylinders. To compensate for this foaming action, experienced locomotive engineers carried the water in the boiler a little low. It took a knowledgeable fireman to know how low to maintain the water in the boiler without uncovering the crown sheet and risking an explosion, particularly when the train was on a downgrade. Early in the twentieth century, roads like the Northern Pacific, Santa Fe, and Southern Pacific introduced water-treatment plants, even though they represented expensive initial and long-term operating costs. Some companies relied on chemical compounds that were placed in either water towers or tender tanks.

It would be during the early part of the twentieth century that railroads introduced a clever method of "taking on" water. After all, this was a practical problem since steam locomotives needed water more often than coal or oil. Inspired by successful efforts in Great Britain and on the PRR as early as the 1870s, some carriers that operated high-speed passenger trains, most notably the NYC and the PRR, thought that the best way to eliminate water stops was by loading water in the tender while the train was *moving*. The collection system involved a movable scoop on the bottom of the tender that could be lowered into a long 6-inch water-filled track pan or pit, usually about a quarter-mile in length that allowed intake while the train was in motion. Railroads were not alone with this concept. Early on aviation companies developed "gas stations in the air," namely, ways to refuel while airborne. By the 1920s in-flight refueling had developed, although for years it has remained a highly risky job.

ELECTRIC LOCOMOTIVES

Even before the appearance of Lima's 2-8-2s and the other examples of modern steam locomotion, another power type, the electric engine, seemed to hold promise for passenger and freight assignments. Differing from the steam locomotive, the evolution of the "electric" was long and not always steady. In the early 1840s Scottish inventor Robert Davidson produced a rudimentary battery-powered contraption that he named *Galvani* in honor of Luigi Galvani, the discoverer of current electricity. In tests conducted by Davidson on the Edinburgh & Glasgow Railway, this vehicle and a small railway carriage moved along at a mere 4 miles per hour. Speed, and even durability, never became hallmarks of this battery approach to electric traction.

This earliest recorded experiment with a full-size electric locomotive failed to bring about the electric railway era. What seemed lacking were positive answers to industrialist Henry Ford's famed comments that innovative success is contingent upon three questions: "Is it needed? Is it practical? Is it commercial?" Finally, by the 1880s, experiments suggested that the electric locomotive might be financially viable at least initially for a niche market. The common wisdom held that the potential of battery power was limited and that success would come from a system that used electric generators, located in a central power source, with overhead wires or third "power" rails to distribute current to rail vehicles.

One accomplishment of note involved a New Jersey electrical manufacturer and inventor, Leo Daft. In 1885 his experimental 9-ton locomotive *Benjamin Franklin*, with its novel direct current (DC) electric motor-mounting and drive arrangement, propelled a commuter train that reached a speed of 30 miles per hour during a test run over about 2 miles of the Ninth Avenue Elevated Railway in New York City.

Yet it would not be Leo Daft who would become the "father of the electric railway" but Frank Sprague, a graduate of the U.S. Naval Academy and onetime associate of the electric wizard Thomas A. Edison. In 1887 and 1888 Sprague's successful electrification of the Richmond (Virginia) Street Railway ushered in the electric railroad era, first with street railways and then with interurban carriers.

The keys to the Sprague triumph in Richmond involved three essentials: an electromechanically efficient DC motor of 500 volts, a nose-suspended method of traction-motor mounting, and the multiple-unit control system. The emerging electric railway industry seemed most impressed that Sprague had solved the nagging problem of how to locate correctly the motor. Heretofore, other inventors had either placed the motor directly on the truck but unfortunately, over time, vibrations proved ruinous, or they had installed the motor in the car body where it became impossible to make a reliable connection with the axle.

Electricity for transportation involved more than using it to replace animal- and cable-powered street railways and applying it to intercity trolleys or interurbans, once a means had been determined to make feasible *long-distance* electric railways. (In 1896 a system of three-phase alternative current [AC] transmission was perfected that significantly reduced voltage losses. Refinements continued, most notably the efficient change by rotary converters of AC to DC power.) Significantly, in the 1890s electrically powered trains, which operated over steam railroads, would also begin.

Arguably the first major steam road electrification success story involved the Howard Street Tunnel of the Baltimore & Ohio Railroad (B&O) in Baltimore, Maryland. "When the . . . tunnel is completed and the trains are hauled back and forth by the subtle power of electricity," opined the *Maryland* (Baltimore) *Journal* of May 28, 1892, "it will be one of the most wonderful events of the world. If it is a success, and there seems no doubt, it will completely revolutionize railroad power and be a great boon to all travelers." In its early assessment of the project, the Baltimore newspaper was correct. Three years later when the Howard Street Tunnel opened, B&O electric locomotive No. 1, one of three engines engineered and built by the General Electric Company (GE), easily handled freight and passenger trains over the new construction. In one high-speed test, the GE product reached a sustained speed of 60 miles per hour. This piece of motive power, which consisted of two semipermanently coupled single-trucks units, employed 360-horsepower DC motors that were positioned on each axle of its four-wheel truck sets. No. 1 drew its 600-volt DC power through a pantograph from an inverted metal-trough apparatus mounted above the track. A bright future for the electric locomotive seemed to be in the offing.

As the twentieth century began, the electrical and railway worlds became abuzz with various breakthroughs in the field of electric traction. Since standardization in power systems had become prevalent, companies had choices. Generally speaking, there was the option between low-voltage DC or high-voltage three-phase AC, what contemporaries called the "battle of the currents." Smaller operations usually preferred the former because the latter required three separate sources of power—two overhead wires and the rails—and the use of induction motors. Furthermore, AC locomotives had a constant power consumption and could be controlled in speed only by efficiency-loss regulation. Yet, power on AC systems could be transmitted long distances without significant loss of voltage, and locomotive motors lent themselves to "regenerative braking." This meant that they were converted to generators on downgrades to feed power back into the overhead line. Three-phase electrification was popular with the electrification of mountain steam railroads. It occurred in the European Alps, the first Cascade Tunnel electrification project of the Great Northern Railway, and portions of the main line of the New York, New Haven & Hartford Railroad. Freight and passenger units on interurbans, however, were commonly propelled by systems of low-voltage DC. Yet, suppliers had an impact on what power was used; General Electric preferred DC, and Westinghouse backed AC.

The value of electrification and the electric locomotive became apparent to the railway world. In a fancy publication, *Results of Electrification*, produced in 1915 by Westinghouse's Heavy Traction Department, the message was clear: "The object of this electrification is to obtain an increase in the capacity of the existing tracks, by increased speed, by hauling heavier trains, and by eliminating congestion caused by delays for coal and water." Added the text writer, "Another object is the reduction of the over-all operating expenses, in maintenance, crews, and fuel" (*Westinghouse Electric Railway Transportation*, 1978). Westinghouse took pride with its recent Elkhorn Grade Electrification project on 27 miles of main line trackage of the Norfolk & Western Railway in the rugged coalfield region of southern West Virginia. When steam ruled, the railroad required three Mallet locomotives, equipped with superheaters and mechanical stokers, to handle a 3,250-ton coal train at 7 miles per hour on the dominant 2 percent grade. With electrification, however, two Baldwin-Westinghouse semipermanently coupled box-cab units, drawing "juice" from an 11,000-volt, 25-cycle AC system, handled the same tonnage trains at double the previous speed. Everyone applauded the performance of these giant locomotives that were equipped with constant-speed, three-phase induction motors. Yet, it became obvious that electrification was expensive (the Elkhorn Grade project cost about $3 million) and was usually undertaken if there were convincing technical, economic, or other reasons.

Interestingly, engineers at this time contemplated what turned out to be a dead-end type of locomotive technology, namely, a hybrid coal-fired steam-electric locomotive. What appealed to enthusiasts were its price and the availability of parts. Since electrification involved expensive fixed facilities, overall costs could be reduced if some type of steam-electric system could be devised, employing some form of motor, for example, a single-phase or three-phase one. Since it was a steam engine that would produce electricity, a boiler, running gears, and a variety of parts would continue to be used, which appealed to steam locomotive suppliers. In 1909 the North British Locomotive Company, located in Glasgow, Scotland, built the prototype of what it called the Electro-Turbo-Loco. Using a steam turbine, the locomotive was a 1000kW DC generation plant on wheels, complete with vacuum condenser, cooling water circuit, air heaters, and turbine generator. High production costs, technical problems, and other drawbacks, however, never permitted this locomotive to enter revenue service in either the United Kingdom or elsewhere. And later experiments to create a dual steam-electric locomotive never led to commercial success.

INTERNAL COMBUSTION LOCOMOTIVES

After the turn of the twentieth century, locomotive engineers began to give serious attention to having internal-combustion engines power railway vehicles. Early and impressive breakthroughs occurred in the area of self-propelled railcars, a technology that later would be exploited by makers of passenger and freight locomotives.

In the late nineteenth and early twentieth centuries, the desire on the part of railroads to find an inexpensive alternative to standard steam trains for lightly trafficked lines led to various experiments. The objective was to build a low-cost steam-propelled unit that could accommodate passengers along with shipments of mail and express. The most serious endeavor involved the Kobusch-Wagenhals Steam Car, a commercially unsuccessful prototype that the giant St. Louis Car Company of St. Louis, Missouri, built in 1906. Its power source was relatively impressive, consisting of a duplex horizontal engine that drove the forward truck through gearing with side rods connecting both front and rear drivers. The boiler was a husky marine-type, water-tube unit with a working pressure of 250 pounds per square inch (psi). Oil was fed to the burners under air pressure, and the entire engine section took up a modest 10 square feet of car capacity. But this steam contraption and others like it had serious drawbacks. Probably the most severe limitations involved the weight of the engine and boiler, the space required to carry fuel and water, the upkeep in managing the recurring clogging, the scaling and corroding of the boiler and pipes, and the constant attention required while operating under high pressure to prevent a boiler explosion.

The experimentation with steam for self-propelled cars was hardly a surprise. After all, railroad officials were "steam men," and as such they thought only in terms of external-combustion engines. But William R. McKeen Jr., superintendent of motive power and machinery for the Union Pacific Railroad (UP) in Omaha, Nebraska, would do much to change that thinking. Encouraged by the enlightened head of the railroad, Edward H. Harriman, in 1904–1905 McKeen experimented with both an appropriate body design and a workable combination of gasoline engine, transmission, chain drive, and other mechanical features.

The result was the "knife-nosed" or "windsplinter" units with their distinctive pointed nose, rounded end, center-loading doors, and porthole windows. Some thought that the vehicle resembled a submarine on wheels. These McKeen cars were cheaper (about 12 cents per mile) and more dependable than battery-powered vehicles and more flexible than steam locomotives. And they could establish competitive operating speeds. With two gear ratios, the motorman (one of only two crew members) could select either "position one," which meant full engine power designed for starting, climbing, or pulling a passenger or express trailer, or "position two," used for high-speed running at 40 to 60 miles per hour. While passengers might not fathom the gear ratios, they surely sensed (correctly) that the internal air system was of new design. Fresh air entered at the front of the car and exhausted through ventilators on the roof. The complete air change cycle took only 4 minutes. The interior lighting was also unique: oval light containers with their opalescent faces provided bright but indirect illumination.

A prototype and improved versions of the McKeen car led to the modern self-propelled gasoline rail motor vehicle. By the eve of World War I these cars, produced by UP subsidiary McKeen Motor Car Company, based in Omaha, had established themselves on the UP and other carriers, especially smaller roads. These 200–300-horsepower all-steel vehicles scampered down tracks throughout the nation and overseas as well. As late as the 1930s some of the 150 units constructed by McKeen were still in revenue service. Indeed, several McKeen cars, albeit with more modern engines, worked branch lines and secondary main lines of the Chicago Great Western until 1950. "The sleek contours with porthole windows, the gasoline engine mounted in front, and other features anticipated the streamline concept," rightly concluded UP historian Maury Klein (Klein 1989, 298). Yet the point should be made that the "streamlining" was all wrong; the front should have been rounded or shovel-nosed and not pointed.

Also of significance, other manufacturers, most notably General Electric, entered the motor car field. It would be GE that made some major refinements, including an electric rather than a mechanical transmission and, for these "second-generation" units, gas-electric propulsion that meant the gasoline engine provided energy to drive a generator that in turn supplied power to traction motors.

As the twentieth century progressed, an important variation to the McKeen and General Electric self-propelled gasoline cars developed. This would be the "railbus," created principally by the J. G. Brill Company of Philadelphia, Pennsylvania, a worldwide maker of electric railway passenger cars and equipment. In 1921, working with Mack Truck, Inc., the Brill firm mounted a 2.5-ton Mack bus chassis and gasoline engine on flanged wheels. The 29-foot vehicle, which seated thirty-one passengers, could regularly move at 30 to 35 miles per hour, even on rough track, although the ride was not particularly comfortable. Later Brill decided to build its own gas railcars, engine and all. Its "Model 55" railcar, introduced in 1923, was 42.5 feet in length, had more seats and cargo space, and was powered by a four-cylinder, 68-horsepower engine. Like the larger gasoline and gasoline-electric rail cars, these railbuses, which varied slightly over time, often operated for decades, usually on shortlines and smaller foreign roads, especially in Latin America.

The major technology replacement that involved railroad motive power featured the diesel engine. It would be this mechanical device that by the 1960s would fundamentally reconfigure the physical landscape of railroads, redefine the role of labor, and relegate steam locomotives to scrap yards and museums. Similarly, the diesel would have a profound impact on other forms of transportation, including ships, boats, and later motor trucks. But in some instances railroads were painfully slow in embracing diesel-powered locomotives, in part because of their enormous capital expenditures already invested in steam locomotives, repair shops, and support facilities. And scores of roads continued to have major investments in coal mines, and they and other carriers also did not want to offend one of the largest single users of the rails, bituminous coal producers.

What became the diesel revolution dates from the closing years of the nineteenth century. In 1893 the brilliant German engineer and inventor Rudolf Christian Karl Diesel published his major opus, *Theory and Construction of a Rational Heat Engine*, and obtained his first patents. Diesel contended that the steam engine was an exceptionally poor thermal machine; for example, the average efficiency of a stationary steam engine was merely 8 to 10 percent, while a steam locomotive had a thermal efficiency of just 6 percent.

Five years following publication of *Theory and Construction* and after considerable testing and adjusting, a commercial diesel engine appeared. Yet there were problems, largely because the science was somewhat ahead of the engineering; builders had to cope with poor metal and frequently crude manufacturing techniques. By the eve of World War I, however, dependable and efficient diesel engines could be found as stationary power sources in Europe and North America, and they even powered British and German submarines. Even though European firms after 1913 had produced dozens of diesel locomotives, it would take more than a score of years before the diesel found practical application in American railroading. The gasoline-electric approach looked promising, yet most railroad motive power tasks required more power than this type of engine could economically produce.

The diesel, nevertheless, offered an alternative possibility, but there existed at least two major concerns. Early diesel engines were heavy (weight per horsepower) when compared to gasoline engines, although by the early 1930s lightweight versions became available, largely through the creative efforts of the Winton Engine Company, a subsidiary of General Motors (GM). Another problem was also solved, namely, a means to transmit the power of the constant-speed diesel engine to the varying-speed torque requirements of a railroad locomotive. The solution, which came in the late 1920s, involved employment of an electric transmission, proving to be both reliable and flexible.

Even before the Winton Engine Company produced a competitive diesel engine, there were prototype locomotives. In 1924 GE and Ingersoll-Rand teamed up to produce a 300-horsepower diesel-electric locomotive, which they demonstrated on several Eastern roads. American Locomotive (Alco) then joined these two manufacturers to create a line of 300- and 600-horsepower diesel-electric locomotives. This corporate partnership rightfully can lay claim to production of the earliest commercially successful diesel-electric locomotives.

It is generally acknowledged that the Central Railroad of New Jersey (Jersey Central), with its GE-Ingersoll-Rand-Alco 60-ton switcher, No. 1000, became the first railroad to have diesel-electric in regular, revenue service. The date was October 22, 1925. Company officials immediately discovered that No. 1000, with its six-cylinder inline 10×12 Ingersoll-Rand diesel engine driving through a GE transmission in an Alco body, was an ideal switcher, even though it was relatively heavy and slow. Being most powerful in the low speeds that switching required, the locomotive easily worked twenty-four hours a day, and performed well at its assigned tasks, namely, functioning in a tiny, crowded yard with extremely sharp curves. Moreover, No. 1000's diesel engine, which generated electricity that provided tractive power, required a minimum of servicing and did not belch

The diesel-electric locomotive did not directly follow the steam locomotive. Rather, builders experimented with fuels and engines. In the 1920s the Ingersoll-Rand Company designed and built an oil-electric locomotive for the Chicago & North Western to use for switching operations in the Chicago area. Courtesy of the author.

out clouds of smoke, and it consumed the cheapest of fuel oil. By 1931 thirty-one more of these units had begun their careers of shunting cars.

As technological improvements continued, everyone interested in perfecting diesel-electric locomotives came to believe that this particular replacement locomotive had already proven itself in switching duties. No one really challenged the notion that the diesel switcher outstripped the steam locomotive in every comparison. Even though the Great Depression, which sent scores of carriers into bankruptcy, dampened sales of all kinds of locomotives, nevertheless, by 1936 there were 195 diesel switchers toiling in American and Canadian yards. Even crusty steam men began to concede that the future of the steam switch engine was bleak.

In the 1930s individuals on the cutting edge of railroad locomotive technology saw wonderful possibilities for producing diesel-powered *passenger* trains. Differing dramatically from the bulky and slow-moving diesel switchers, they realized that the major passenger-carrying railroads demanded motive power that was dependable and safe, and they universally placed a premium on fast schedules. This optimism grew from research at General Motors following the perfection of a two-cycle, eight-cylinder diesel engine. Moreover, metallurgists had developed new high-strength steel and aluminum alloys that possessed a relationship of weight to strength making such a power plant light enough to mount on a locomotive frame and powerful enough to pull a train. Then, too, breakthroughs were occurring in the field of aerodynamics. Aircraft builders were taking advantage of "wind-tunnel" testing, finding appropriate designs to reduce wind resistance through streamlining. Wind-tunnel devices were also beginning to appear on college campuses, including the Massachusetts Institute of Technology in Cambridge and the University of Michigan in Ann Arbor.

The most sensational diesel-electric passenger train was about to be born. In the early 1930s Ralph Budd, the dynamic president of the CB&Q, realized that his Midwestern carrier was feeling the effects of growing automobile competition. The company and other railroads had to deal with a domestic motorcar industry that produced a record 4.5 million passenger vehicles in 1929 alone. Furthermore, carriers needed to confront a public that was developing the attitude that railroads were "tired" and not wholly progressive. So Budd responded. In June 1933, the CB&Q placed an order with a Philadelphia auto-body manufacturer, the Edward G. Budd Company (whose head was a distant relative of the Burlington executive), for a high-speed, lightweight, streamlined, three-car passenger train that would cost an estimated $200,000. Specifically, this seventy-passenger, 196-foot train set would have three distinct units: "The first car will contain the motor, baggage and mail compartment; second car, baggage and express, with a buffet and smoking compartment in the rear to seat 19 passengers; third car will be entirely devoted to seating space with a capacity of 51, which includes 12 parlor chairs" (Overton 1965, 395). Burlington's Budd named it the *Zephyr*, after the Greek god of the west wind, Zephyrus. This was an appropriate selection, for Zephyrus represented renaissance and that was exactly what Budd expected of the diesel-powered passenger train for these depressed times.

The Budd Company did not disappoint Ralph Budd or the CB&Q. When the *Zephyr* rolled out of the Budd Company shops in April 1934 with its GM-built diesel engine, it quickly began trial tests and started an extended national tour, attracting large and enthusiastic audiences. But the big day for this historic diesel-electric passenger train came on May 26, 1934, when it made its unprecedented Denver, Colorado–Chicago, Illinois, nonstop run of 1,015.4 miles in an amazing 13 hours, 4 minutes, and 58 seconds at an average speed of 77.61 miles per hour. (Its highest speed reached 112.5 miles per hour.) As the *Pioneer Zephyr*, the United States' first diesel-powered streamliner went on to travel 3.2 million miles and carry more than a million passengers before its final retirement in 1960 to an exhibit hall at Chicago's Museum of Science and Industry. It did not take long before the public rendered the *Zephyr* and other similar trains synonymous

In the 1930s the Chicago, Burlington & Quincy Railroad took joy in promoting its growing fleet of streamlined diesel-electric trains. The company noted on this advertising card that on October 23, 1936, one of the new *Denver Zephyrs* set a world's speed record for long-distance nonstop runs by sprinting from Chicago to Denver, a distance of 1,017 miles, in slightly more than 12 hours. Courtesy of the author.

with color and speed and comfort and cleanliness. It should be remembered, too, that the *Zephyr* emerged as a symbol of hope at a time when America was scourged by depression and doubt.

Even though the CB&Q's entry into the field of diesel-electric powered passenger trains grabbed the national (even the world) spotlight in 1934, another version of the modern streamliner made its debut. Early that year the Pullman-Standard Company produced for the Union Pacific, that trailblazer in gasoline motor cars, what was called the *M-10000*, a golden-brown and canary-yellow articulated streamliner. "Little Zip," as UP shopmen called the $200,000 train (the railroad officially dubbed it "Tomorrow's Train Today"), consisted of three cars made of an aluminum alloy that weighted a modest 85 tons. (With their emphasis on weight reduction, both the CB&Q and UP trains had a somewhat smaller cross-section—both height and width—than previous or later trains.) The *M-10000* accommodated 116 passengers and crew together with 25,000 pounds of baggage and mail. The power plant featured a 600-horsepower Winton V-12 electric engine made of wrought steel. A state-of-the-art Westinghouse generator ran each of the two GE traction motors that were mounted on the front trucks. This mechanical system allowed the *M-10000* on its maiden voyage in October 1934 to seemingly fly across the vast territory between Los Angeles and Chicago, making the trip in less than 39 hours, clipping 6 hours off the fastest previous run between these two cities.

Although both the *Zephyr* and *M-10000* were fast and flashy, there were differences. Unlike the *Zephyr*, the *M-10000* (later renamed the *City of Salina*) not only sported riveted aluminum skin but also did not consume diesel fuel. Its engine, with spark ignition, ran on a petroleum distillate, which by then the UP had found satisfactory for propelling its fleet of rail motor cars. Soon, though, when the UP took delivery of a similar streamliner, *M-10001*, its power plant contained a diesel engine; there were no spark plugs or distillate fuel for this train that became the *City of Portland*.

Within a year after the introduction of the *Zephyr* and *M-10000* and *M-10001*, additional diesel-powered streamliners entered revenue service. The New York, New Haven & Hartford, for one, inaugurated its *Comet*. Built by Goodyear-Zeppelin, the three-unit lightweight stainless steel train with its Westinghouse six-cylinder engine packed 800 horsepower, carried 160 passengers, and became the only bidirectional streamliner of the 1930s. (This latter feature meant that with its either-way equipment, terminal turnarounds could be avoided.) And little Gulf, Mobile & Northern scored a major "first" with its two American Car & Foundry (ACF) -built and Alco-made six-cylinder inline diesel-powered *Rebels*, the nation's first nonarticulated (or non-fixed-car) streamliners. The diesel-power streamlined era in railroading

Just as the Chicago, Burlington & Quincy ballyhooed its fleet of "America's Distinctive Trains," the Union Pacific Railroad, the other pioneer in streamlined passenger travel, made much of its "City" trains. In 1936 the articulated *City of Portland* heads for the West Coast on the outskirts of Omaha, Nebraska. Some observers thought this unique train resembled an angry caterpillar. Courtesy of the author.

had arrived; steam-powered passenger express trains or "limiteds" were doomed.

Yet it would be misleading to suggest that streamliner technology moved in a steady, straight and unbroken fashion. As with most technologies, "dead-ends" occurred; a particular set of drawings or an actual prototype never became anything more. This would be the case of the *Clark Autotram*. Engineers at the Battle Creek, Michigan–based Clark Equipment Company, working closely with representatives from the Aluminum Company of America (Alcoa), created what some believe to be the "first" American-constructed streamliner. A 1933 promotional brochure stated that the *Autotram* was "built for swift, frequent passenger service, [and] embodies many innovations in design." Added the text writer, "It is fully streamlined and powered with a sixteen-cylinder automotive type engine. The double-pane, shatter-proof windows are stationary as the car is equipped with a forced ventilation system." Allegedly, the *Autotram*, fabricated with strong Alcoa-produced "duralumin" alloys, could whisk its forty passengers down the track at speeds approaching 100 miles per hour. But the Clark company's proud and hopeful message that the *Autotram* "represents a distinct advance in the progress of rail car construction and makes available a new phase of railroad passenger service embodying frequent, high-speed schedules and the latest developments in comfort and safety" went nowhere (*Autotram* 1933). Although prominently displayed at the Century of Progress Exposition in Chicago in 1933 and 1934, the *Autotram* found no industry support probably because of its cost and the ever-more crippling national depression. The Clark company then wisely concentrated on manufacturing forklifts and other industrial equipment.

As diesel-electric locomotives moved successfully into daily passenger service, attention turned to their development as over-the-road freight units. In the later part of the 1930s the Atchison, Topeka & Santa Fe Railroad, working closely with the Electro-Motive Division of General Motors (the former Electro-Motive Corporation), became the first railroad in the United States to operate a diesel-electric freight engine. In early 1938 the 5,400-horsepower, four-unit blue and yellow locomotive, with its sixty-four cylinders, fourteen generators, and sixteen motors, arrived on the property and immediately dazzled the operating department. The units moved the assigned tonnage of sixty-six carloads of heavy freight from Kansas City, Missouri, to Los Angeles, California, with speed, efficiency, and economy. Specifically, the diesel freight went through with only five brief fuel stops; normally, it took nine individual steam locomotives to make the trip, with thirty-five stops for fuel and water. This husky GM locomotive set, moreover, had the advantage of being split into two 2,700-horsepower units. Unfortunately, the economic dislocations caused by World War II prevented many more of these road diesels from entering service. But once peace returned, that would change dramatically when this type of modern freight power altered the face of American railroading.

STEEL CAR ERA

While the introduction of shiny diesel-electric passenger streamliners became newsworthy events, the advent of nonstreamlined all-metal passenger and freight cars was much less heralded. Perhaps rolling stock, certainly pieces that handled freight, lacked much public interest. Moreover, they were really not wonderful examples of engineering and technological change. Yet these improved cars were important to the carriers and arguably, too, to riders and shippers.

John H. White Jr., the premier student of the American railway rolling stock, has noted that for passenger cars the period from the mid-1840s to the dawn of the twentieth century was characterized by extended experimentation with metal-building components, but the thirty or so years following 1901 turned out to be the era of metal cars, what might be considered the standard or heavyweight phase. Then it would be the *Zephyr* and *M-10000* that ushered in the time of lightweight construction.

Surprisingly, there existed cogent reasons for railroads *not* to embrace all-metal construction. Even though for decades inventors, including Cincinnati sheet-metal fabricator Edward Robbins, had demonstrated that such passenger equipment was practical, industry leaders rightly contended that a metal car could not really carry any more passengers than existing equipment. Moreover, such pieces of rolling stock, being heavier and more costly, would increase operating expenses. For some time, however, these executives accepted the practice of reinforcing wood cars with steel members, but wood continued to dominate.

By the early years of the twentieth century, safety became the leading reason for railroads to accept metal replacement technology. Like automatic car couplers, this was a technology that was inherently "accident saving." With increased speeds, additional traffic, and faster locomotives, train derailments that involved wooden equipment resulted in greater loss of life. When telescoped into a steel locomotive tender or other steel piece of rolling stock, wooden coaches shattered. And gas-light illumination and coal-burning stoves could easy set the splintered cars ablaze, turning them into pyres that roasted victims who would otherwise survive the impact. When the *Continental Limited*, the flagship train operated by the Wabash Railroad, collided head-on with another train near Sand Creek, Michigan, on November 27, 1901, heavy Pullman cars crushed the flimsy wooden coaches and an uncontrollable fire ensued, killing perhaps 100 passengers and crew members. This deadly accident and others like it enraged the public, promoting demands for steel cars and the end to any open flames in car interiors.

It was the fear of fire that prompted the Interborough Rapid Transit (IRT) in New York City to order from the ACF 300 all-steel fireproof coaches. To reduce weight, the manufacturer, for example, opted for pressed shapes and aluminum panes and fittings for the interior. These cars entered service in 1904, and were the first of their kind in the United States. Also, a number of steam railroads, most notably the PRR, which had built a prototype car for the IRT, turned to all-steel designs. Although some carriers and commercial builders opted for only metal sheathing of wooden coaches for cost and weight considerations, there was universal acceptance of steel underframes. And quickly the design that dominated was known as the "fishbelly," namely, a massive center sill made of two trapezoidal steel plates. These fishbelly units worked well; they were incredibly strong and easy to assemble, and, unlike wooden floor frames, they fully supported the sides and tops. Of course, steel passenger cars were hardly indestructible, but they were far safer than rolling stock of the wooden era. Travelers, of course, benefited from the changeover from wood to steel, even though substantial numbers of wooden cars remained in service for years, usually on branch lines and shortlines.

When the CB&Q and the Union Pacific ushered in the streamliner era, there was a wave of experimentation with various types of car-body construction before wartime demands halted passenger-car production.

The Budd Company built shot-welded stainless steel cars with truss-side frames; ACF constructed riveted girder-type cars of both aluminum and low-alloy, high-tensile steel; the Pressed Steel Company fabricated lightweight riveted steel cars; and Pullman manufactured several varieties, including aluminum girder types, low-alloy-steel truss frames with fluted stainless-steel coverings, and low-alloy-steel girder-types with either flat, painted sides or fluted stainless steel applied over the girder sheet. Modernity had fully arrived for the passenger car and builders, and carriers happily touted their new acquisitions.

The public did not fuss about the construction of freight cars, although they assumed that cars would protect their contents. As with passenger equipment, there was a revolution in materials and construction for these pieces of rolling stock. As with the steam locomotive, early experiments in England and in other European countries led to important advances in freight-car-building technologies. In the late 1880s, for example, the Iron Car Company of Great Britain delivered a large order of all-metal coal cars and flatcars to the Furness Railway in Scotland. Much earlier the Prussian State Railways in Germany had employed metal cars; after 1870 they began to enter service. Then in the late 1880s, small numbers of large-capacity steel flatcars appeared on the rosters of American carriers. About this time, the PRR, for one, built several "special service cars" to carry heavy naval cannons and coils of wire rope.

In the 1890s the metal freight car was coming of age, albeit somewhat slowly. By 1900, however, production of this equipment had increased significantly: 29,800 manufactured as compared to 112,000 of wood construction. Five years later the figures stood at 75,000 steel and steel-framed versus 90,400 wood cars, and in 1910 steel had dramatically surpassed wood: 162,600 to 13,700.

Although the safety issue was a driving force in explaining why passenger-car builders opted for steel, different factors account for the steel freight-car revolution. Yet some of the same reasons explain why wood had largely disappeared in the construction of almost every piece of rolling stock by the era of World War I. Of course, some wood would continue to be used, for example, for planking floors in steel boxcars and flatcars.

The changing nature of car-building personnel had a pronounced impact on the popularity of steel and other metal products. Individuals who decided about the choice of construction materials were often of a generation that had collegiate or technical-school training. Their predecessors likely had been apprentices in cabinet shops, sawmills, or wood-working plants or had other important ties to such activities. The new breed of car makers had been exposed to trade journals, technical papers, textbooks, and other sources that described and analyzed steel as an essential component in bridges, ships, and skyscrapers. These men knew the strengths and limitations of building components, realizing that the age of stone and wood was rapidly passing.

Steel became even more attractive for freight-car manufacturing with the introduction of pneumatic tools. By the early twentieth century an array of power equipment, including chisels, drills, and riveting hammers, was coming into widespread use. Workers soon felt comfortable with these replacement tools, and managers happily found these essential products for steel fabrication declining considerably in price. The presence of better tools and their growing demand in other industries do much to explain their attractiveness in terms of both quality and cost.

Additional cost considerations came into play. As result of the enormous impact of the Pittsburgh-based Carnegie Steel Company, which became universally praised for its quality control, efficiency, and mass production, prices for steel and steel-related products had fallen, often dramatically. For example, by the early 1890s rolled steel beams sold for about $40.00 per ton, although not long before they had commanded about $70.00 per ton. High-strength wood products, particularly white oak and Southern yellow pine, were becoming more difficult to find and hence more expensive.

Railroad executives came to realize that steel offered a practical way to reduce their operating costs. Since freight traffic soared after the return of widespread national prosperity in 1898, following nearly six years of severe depression, the steel freight car offered a way to increase tonnage per unit and to reduce operating expenses. Simple mathematics demonstrated the superiority of this replacement technology. To have a train with 1.5 tons of freight capacity, heretofore it required fifty 30-ton wooden cars; but only thirty 50-ton steel cars could carry the same cargoes. Moreover, the collective weight of the wood equipment (including metal trucks) exceeded 800 tons, but the smaller number of steel cars weighed about 300 tons less. Fewer cars meant fewer locomotives, operating crews, and operational hassles. Efficiencies and savings resulted. And officials discovered other cost advantages, including repairs. Switching crews became less careful when handling cars with automatic Janney-type couplers than they had been when link-and-pin couplers reigned. A coupling crash might damage a car of wooden construction, whereas a car made out of steel usually suffered little or no damage at all.

OTHER PASSENGER CAR BETTERMENTS

Long before the first streamliners thrilled the public, important, long-lasting technological improvements were occurring in passenger rolling stock. For one thing, dramatic changes were taking place with illumination. Although gas lights were the best methods of artificial illustration for much of the nineteenth century, electricity offered a brighter alternative. Gas lights, which were far superior to candles and oil, had a tendency to explode when not handled carefully, and so inventors kept looking for a better light source. The incandescent electric bulb seemed to be the answer; after all, it did away with open flame, smoke, and burning gases. As early as 1881, not long after the introduction of commercial electricity, a few railroad cars were outfitted with electric lamps. The place, though, was England and not the United States. The London, Brighton & South Coast Railway led the way by installing a dozen Edison lamps aboard a sleeping car. Thirty-two wet-cell batteries furnished the current that maintenance workers nightly recharged. In 1882 the PRR tried the British experiment. By the end of the 1880s that company operated a modest number of cars with electric lights, although steam-driven generators ("head-end plants"), located in baggage cars, customarily provided the current. Still, those railroads that offered this lighting source usually relied on rechargeable batteries even though they tended to be expensive, heavy, fragile, and short-lived.

The progress of replacing gas with electric interior lamps was relatively slow in the United States. In 1898, for example, only a few hundred cars featured electric illumination. In Europe, on the other hand, an estimated 6,000 cars were so equipped. American railroads believed that the cost, the trouble, and also the workability of manufactured gas did not warrant the changeover. Admittedly, some companies liked the publicity value of saying that their top-notch limiteds featured electric lights. The small Minneapolis & St. Louis–Iowa Central system, for one, took considerable pride to proclaim on the cover of its May 1910 public timetable that it offered "ELECTRIC LIGHTED TRAINS." This pronouncement implied that this Midwestern road was up to date and that patrons need not to worry about the potential danger of open-flame lighting.

Even though the practical solution to electrification of passenger cars dates from the 1880s, American inventor Morris Moskowitz is commonly credited with the success of this approach. What he did in the 1890s was to advance the technology associated with the "axle-light system," specifically winning several patents that involved an axle-mounted pulley belted to an electrical generator. Early on, though, axle lighting had the shortcoming of not being able to maintain a constant voltage when the generator's rotation fluctuated with the speed of the train. Yet this problem was ultimately solved by employing a specially wound field magnet. It would be the work of United States Light and Heating Company, using Moskowitz's concepts, that led in 1902 to the NYC's premier all-Pullman train, *The Twentieth Century Limited*, to add axle lighting. With improvements and positive railroad responses, the number of electrically lighted cars soared, reaching approximately 15,000 by World War I.

In the formative years of the twentieth century when travelers began to enjoy electric lights in chair, dining, and sleeping cars, they also came to appreciate advances in heating. As open flames for illumination posed a fire hazard, so stoves, the conventional source of heat since the dawn of railroading, likewise created a fire threat. The long-term solution involved some form of steam heating.

The idea of somehow tapping steam from the locomotive came at an early date. But it would not be until the 1860s that the concept became much of a reality, most successfully undertaken on railways in Prussia and Sweden. Then in 1878, New Yorker William Baker demonstrated a promising steam-heating apparatus for cars on the New York (City) Elevated Railway. Baker used pipes with flexible rubber hoses to draw steam from the locomotive (rather than an auxiliary boiler) into the passenger equipment. In this case, the introduction of the Baker Heater had much to do with an increasing demand by management to have a safe, compact, and efficient heat source.

Even though Baker's approach to steam heating worked reasonably well for the New York Elevated, railroad executives found excuses for not adopting the Baker product. At the heart of their arguments was the notion that the locomotive was meant to haul cars and not to heat them. Yet the problem of having safe and dependable car heating did not disappear. It would be Edward Gold, however, who improved on Baker's work, so much so that by the early twentieth century he could claim that his system had been installed on more than 40,000 cars. "NO MORE PASSENGERS BURNED ALIVE: THE DANGEROUS STOVE SUPERSEDED" became the proud advertising cry of Edward E. Gold & Company (White 1978, 394).

The Gold response to the heating conundrum was to follow upon Baker's interest in locomotive steam. In experiments conducted in the early 1880s, again on the New York Elevated, Gold managed to overcome a weakness of the Baker system, namely, the inability of locomotives, with their frequent starts and stops, to supply a constant stream of steam for heating. Gold decided to have the available steam delivered to hot-water heaters in the cars once the train was underway (when the need for maximum steam was temporarily ended). Specially designed radiators, with double-tube reservoirs and an inner pipe that contained salt water, permitted the retention of enough heat to keep a car warm for several hours. In time, Gold and his New York City–based company made modifications and other improvements, offering direct steam, steam vapor, and electric systems. By the 1910s, thanks to Baker and Gold, winter weather and cold snaps no longer plagued most railroad travelers. However, the ubiquitous coal-burning potbelly stove remained in most cabooses until after World War II. Similarly, these pieces of rolling stock continued to use kerosene lamps nearly as long, although by midcentury mechanics might install a "Spicer Drive" generator that allowed the moving wheels to produce electricity for inside illumination.

COMMUNICATIONS AND SIGNALING

When railroaders spoke about "modern" communications throughout much of the latter half of the nineteenth century, they usually meant the electric telegraph. After all, it was the clicking of telegraph sounders that informed divisional dispatchers, depot agents, and operating crews about train movements and other matters of railroad business. By the end of the century, however, some carriers began to use the telephone for communications, including train dispatching. It was a small, high-density, narrow-gauge commuter road, the Boston, Revere Beach & Lynn Railroad, that was probably the first to use telephones for train management. It did so in 1879, only a few years after the introduction of Alexander Graham Bell's grand invention. But seemingly not too many railroad orders and other communications were initially given by voice. The influence of the telegraph remained strong; most routine telephone messages were given by signal bells. Apparently, railroad employees were not ready to accept the exclusive use of the spoken word to transmit orders.

Despite the gratifying results obtained by the Boston, Revere Beach & Lynn, no great rush developed from other railroads to adopt the telephone. Of course, by the 1890s the public likely could make telephone contact between their homes and offices and railroad depots, freight houses, and other facilities. Indeed, with a few exceptions, it would not be until about that time that other railroads started to use telephones with any regularity for train-control work. This reluctance was only partly due to conservatism; the lack of suitable methods for calling outlying stations was an important factor. Then, early in the twentieth century, Bell Laboratories started to work on applications of "telephony" to the specific needs of train dispatching. Because nearly every message concerned the dispatcher, the communication network was in reality a heavily loaded party line. The Bell engineers concentrated on developing some method of signaling that would be selective rather than code ringing. Initially, a device used on telegraph lines, known as the Gill selector, was modified and adapted for this purpose. Improvements followed and in 1916 Bell introduced the "No. 60 Selector," which remained in service with few modifications as long as telephone dispatching continued.

Telephones usage grew rapidly. By 1914 more than 11,000 miles on 172 railroads were dispatched exclusively by telephone. Then in 1925, for the first time, mileage dispatched by telephone exceeded the mileage controlled by telegraph. Even before the major Bell triumph, the superintendent of telegraphs for the Atchison, Topeka & Santa Fe Railroad explained the growing popularity of telephones:

> To the dispatcher time is the one thing that must receive greatest consideration. It may be assumed that a man can utter words with a speed equal to the operation of his mind. But the train dispatcher, in using the telegraph, has always been obliged to transmit his words at a speed but one-tenth his capacity to express himself, and also to receive a reply at the same rate. It means that the dispatcher has been permitted to multiply the time within which he may form his plans of trains operation fully three times. This, of itself, further means that his mental calculations are vastly improved in accuracy and general value. There is the certainty that more movements will be better planned and executed, and that the art of direction of operation of trains will be greatly advanced and the work more safely rendered. (Hay 1974, 59)

It would be the telephone that inspired the always innovative Chicago Great Western Railroad to use Dictaphone equipment to convey information of train movements to dispatchers from places where station personnel were off duty. In 1914 the CGW launched its Dictaphone-telephone network. The plan, which functioned rather well, was ingenious: an agent merely connected the Dictaphone apparatus to an open telephone line before he left his office. During the first test, "the dispatcher, fifty-seven miles away, heard the ringing of the engine bell, the exhaust of the engine and roar of a train passing the station." He also recorded the arrival and departure of a passenger train, "identifying it by the noise of loading milk cans" (Grant 1984, 87). For the next two decades the company, joined by other roads including electric interurbans, would use these devices in this capacity.

Improvements in signaling equipment remained ongoing. During the late nineteenth and the beginning of the twentieth centuries, considerable efforts took place to make trackside semaphores viable at night. Early on, a variety of oil-burning lamps with clear or colored lenses were used, either linked physically to semaphore arms or operated independently. But in the

1880s and for two decades afterwards, serious efforts were made to make the blade itself seeable in the dark. Several schemes evolved, including the use of parabolic reflectors on the axis of the blades and white illuminated screens behind signals so that blades could be seen in silhouette. Then these efforts stopped. Not only were night-viable blade products burdened by their high costs and usually complex features, but also after 1910 significant breakthroughs occurred in the field of optics. It now became possible to use color-light signals both at night and also at distances of several thousand feet in *daylight*. About the same time the PRR introduced its position-light signal, and soon thereafter the B&O inaugurated a color position-light signal. These improvements ended the need for night-running engine crews to be able to see semaphore blades.

With colored lights guiding enginemen day and night and increasingly enhanced signal-blocking systems, train safety improved, much to the satisfaction of everyone: labor, management, the public, and government. Occasionally, it would be state and federal agencies that promoted a particular technology, for example, through the Safety Appliance Act of 1893. Since safety remained a major concern of the ICC, which had sought to protect a broadly defined public interest since its inception in 1887, development after World War I of both centralized traffic control (CTC) and automatic train control (ATC) led the ICC to vigorously endorse these technologies.

In the late 1920s the General Railway Signal Company and its rival the Union Switch & Signal Company did much to make possible CTC. This vastly improved alternative to train control revolutionized dispatching, and remained largely state-of-the-art until high-speed computers and special software made their debut in the 1980s and 1990s. Installed usually on only the busiest sections of track, CTC allowed a divisional dispatcher to run his section of a railroad much as if it were a model-railroad layout. The time-consuming and potentially dangerous system of directing train movement with telegraph- or telephone-initiated paper train orders, which needed to be sent to all trains involved through stations along the line and in some cases could be delivered only by having the train stop so that crew members could sign for the orders, no longer was necessary. Trains under CTC could operate entirely by signal indications telling crew members when to stop, where to take sidings (and when they did, switches were already thrown), and when to start again. Crews no longer needed to get down from their trains unless they wanted to telephone the dispatcher. By this time, some carriers had installed telephone boxes every several miles for emergencies.

As the name implies, CTC operations are centrally controlled. The locus of activity is a switchboard in the dispatcher's office. By moving the appropriate levers, the dispatcher opens and closes passing-track switches, which are thrown by small, powerful motors, and sets the signals that inform trains crews what to do. Lights above the levers indicate how the switches and signals are set and when a train is nearing a particular switch. A failsafe feature makes it impossible for the dispatcher to fix signals incorrectly. Above the lights and levers is a lighted track diagram that shows whenever a particular section of track is occupied by a train and, on most CTC boards, beneath the levers and under glass, is a moving sheet of graph paper on which a pen makes a permanent record of every train movement.

In 1927 the first important CTC installation took place near Toledo, Ohio. The NYC realized that it must improve train movements on about 40 miles of highly congested single track. Management was pleased when CTC made it possible for a single track to handle the volume that a double tracking could provide. And other railroads followed the NYC lead. The Great Depression, however, severely limited installation because of the high costs (about $10,000 per mile in 1930), although during World War II, when freight and passenger traffic soared, CTC operations greatly expanded.

The 1920s also saw the introduction of automatic train control systems. This less expensive technology, technically called General Railway Signal Automatic Two Speed Train Control, offered a means of preventing a train from passing a stop signal or slowing down at a caution signal by taking control away from the engineer and placing it in the hands of a mechanical device that applies brakes automatically. Specifically, this continuous-induction system, in which the track circuit itself was used to actuate induction coils mounted on the locomotive, responded to both signals and track conditions. The ATC system reacted to anything that interrupted the circuit, such as a broken rail or an open switch. While the ICC endorsed the installation of networks of CTC, it enthusiastically backed ATC equipment. In 1922 regulators ordered forty-nine railroads to use ATC on their most heavily traveled lines.

The first railroad to make a major commitment to ATC was the Chicago & North Western Railway (C&NW). Although added in phases, by 1928 the company had its 485-mile main line between Chicago and Omaha, Nebraska, protected. And C&NW officials did much to advertize this added safety feature for its freight and passenger movements. The company likened ATC to "a giant hand or *invisible guardian*." And, as it explained to the public, "Automatic train control protects you in the worst kinds of weather, and materially aids in allowing your train to be brought through on scheduled time. It also is of immense benefit to shippers in bad weather, especially shippers of perishable goods, as much delay is avoided because the engineer can run at the maximum speed allowable in spite of weather conditions" (quoted in Grant 1996, 142). For the C&NW and other roads, this added technological betterment resulted in positive benefits, justifying the initial investment and maintenance costs.

Cab signals were an important byproduct of ATC. For many years after their introduction in the 1920s, these devices appeared in conjunction with ATC systems or by themselves. As they developed, these miniature light signals, mounted in the cab in sight of the engineer, gave a constant picture of track conditions by providing what the next trackside signal would show. By 1950, about 10,000 miles of line featured ATC and 3,500 miles were equipped with cab signaling without the automatic control enhancement.

While only the wealthiest railroads might be expected to installed CTC, cab signals, and the like, every railroad, even the poorest shortline, took advantage of the ongoing revolution of sorts that took place in new line construction, relocation, and repairs. Over the years men wielding shovels and pickaxes and driving horse- or mule-drawn wagons and scrappers yielded to power steam shovels, motorized "cats" (caterpillar-type tractors), and dump trucks. Early in the twentieth century the Ohio Shovel Company in Marion, Ohio, demonstrated the value of the steam-powered shovel when it supplied such state-of-the-art equipment to the construction of the Panama Canal. By the 1920s the Caterpillar Tractor Company of Peoria, Illinois, revolutionized both railroad and public construction, and a variety of contemporary truck manufacturers, led by White Motor Company of Cleveland, Ohio, produced ever-larger and better vehicles.

Diesel-electrics, streamliners, CTC systems, and improved construction equipment demonstrate that in Maturity, the railroad industry continued to make important technological improvements. Yet this does not mean that a certain conservatism did not occur. Some railroad leaders were reluctant to make changes, especially if they were not convinced that the status quo needed to be altered. And then there was that ongoing concern about cost, even during the expansive years of the Railway Age. After the progressive reform triumphs of the Hepburn Act of 1906 and the Mann-Elkins Act of 1910, railroads were losing their power to manage rates. Then in 1916 the Adamson Act made the eight-hour day the basis for determining a day's wage for train crews, giving railroad labor one of its sweetest victories. What these federal statutes meant was that the government had taken over both sides of the ledger, namely income and expenses, creating what historian Albro Martin has called "Enterprise Denied." Moreover, by the 1920s railroads faced the financial sting of increased competition from internal combustion transportation alternatives: automobiles, buses, and trucks. These more dependable vehicles could operate on ever-increasing networks of all-weather roads. Soon, too, commercial aviation began to have an effect. Still, industry officials, like Ralph Budd at the CB&Q, understood that railroads had to attack ossification and to accept, even promote, better replacement technologies.

4

Old Age, 1940–1970

STEAM LOCOMOTIVES

When the American railroad industry reached Old Age in the 1940s, the Age of Steam had not yet ended. Throughout the country, remaining in active service, were thousands of steam locomotives, some of which dated back to the turn of the twentieth century or even earlier. Although scores of diesel-electric locomotives joined the rosters of American carriers during World War II, production, especially of heavy-duty freight road units, was limited because the nation's industrial machine concentrated on producing an array of products to defeat the armed forces of Germany, Italy, and Japan. Throughout the 1930s and 1940s, mechanical engineers and their associates eagerly sought to create bigger and better steam locomotives. Indeed, at the beginning of the Great Depression a veteran Baldwin Locomotive Works executive expressed optimism when he told colleagues at a professional gathering that the steam engine remained "the greatest of all human devices," predicting that it "can be more ably discussed in the year 1980 than at this convention in 1930" (*Railway Age* clipping, n.d.). But that would not happen. By the late 1940s, the steam locomotive would be refined to the edge of its technical limitations.

The Union Pacific Railroad (UP), the powerful, prosperous, and innovative long-distance freight and passenger carrier, emerged as a leader in developing and operating the last giants of the Age of Steam. Like other

Even by the mid-1940s many main line steam engines lacked mechanical coal stokers. In 1946 a fireman for the Frisco Railway uses the old hand-shovel method of feeding the firebox. Unlike crews in a contemporary diesel-electric locomotive, crew members in steam locomotives found their work areas cramped, dirty, and noisy. Courtesy of the author.

carriers, the UP wanted mighty power for its long freight trains. Simply "double-heading" (attaching two or more smaller steam locomotives) would not do; it was usually too labor intensive. In the mid-1920s the UP raised eyebrows in the locomotive world when it introduced two powerful engines: the *Overland* or 8000-class, which featured a 4-10-2 wheel arrangement, and the *Union Pacific* or 9000-class, which sported an extra set of drivers (4-12-2). The latter won the distinction of being the largest nonarticulated steam locomotive ever built and featured the longest (30 feet, 8 inches) rigid wheel base of any iron horse. Observers rightly considered the 9000 class to be a compromise between the draft horse and the racehorse classifications. In 1926 both engine types performed well in extensive testings, and experts believed that the *Overland* and *Union Pacific* might well be the power of the future for heavy-freight and fast-passenger assignments. The company so liked these new iron horses that it initially ordered from the American Locomotive Company (Alco) ten of the former and fifteen of the latter. It did not take long before scores of both types entered its power pool, with the 9000s becoming the more numerous.

In time, however, these super steam locomotives disappointed Union Pacific. Within a decade or so, it became painfully apparent to the mechanical department that these engines, most of all the 9000s, broke down frequently, necessitating expensive maintenance. "The giant 4-12-2 9000-class locomotive . . . promised much but ate up its earnings in the repair shop," commented Maury Klein, the historian of the modern UP (Klein 1989, 372). Still, some employees liked these locomotives. "There was something about its lopsided exhaust that even sounded like raw power," remembered a longtime company executive. "Because it had three cylinders, the 9000 exhausted six times with each revolution, instead of four like conventional engines. Thing of it was, the six exhausts didn't sound alike because it was bigger, and made a louder sound. It was the cadence which was staccato and the emphasis was on the middle cylinder: choo-CHOO-Choo . . . choo-CHOO-Choo . . . choo-CHOO-Choo" (Bailey 1989, 120).

As the hard times of the 1930s started to lift, UP added another monster locomotive type to its extensive motive-power fleet. Known as the *Challenger*, this Alco-built engine featured a 4-6-6-4 articulated design. What impressed the railway world was that this locomotive, when operating over mostly level track, could wheel a 100-car freight train at 100 miles per hour. What was really an adaptation of the 2-6-6-4, the *Challenger* became well recognized for its stability at high speeds, making it one of the most popular simple-articulated engines ever built. By 1944 more than 200 of these oil-burning, high-pressure behemoths had been produced for both eastern and western railroads, although the UP claimed the most, a whopping 105.

It would also be the Union Pacific that became the sole owner of what was truly the last of the steam giants, the *Big Boy*, a true Sampson of the rails. The name fit: this 4-8-8-4 articulated locomotive, which first appeared on the property at the start of World War II, held the distinction of being the largest steam locomotive of any type ever built. No one denied that *Big Boy* was the biggest of all. Designed for heavy freight service, even on the most challenging sections of the main line, and engineered to produce maximum power output continuously at 70 miles per hour, its engine weighed a staggering 627,000 pounds and the weight of its tender (two-thirds loaded) reached nearly 350,000 pounds. The *Big Boy* was 132 feet long and carried 28 tons of coal and 25,000 gallons of water. If a firemen were to hand stoke its massive firebox, he would likely produce only enough steam to blow its chime whistle; a mechanical stoker with an Archimedes' screw (a spiral tube coiled about a large shaft) kept ample coal flowing into the firebox. And *Big Boy* was capable of evaporating per hour more than 50 tons of water, or over 12,000 gallons! No wonder these rugged and mighty 7,000-horsepower engines could handle the toughest grades, including rugged Sherman Hill in Wyoming, without the use of "helper" steam locomotives, and could pull a 5.5-mile train on level track.

During the final years of steam locomotive development, the Pennsylvania Railroad (PRR) became an eastern counterpart of UP. The PRR wanted to add the latest and best steam motive power, wishing to supersede its fleet of aged 4-6-2s and 4-8-2s. In 1939 the PRR startled the railroad industry with its revolutionary 6-4-4-6 or Class *S1* duplex-type passenger locomotive. The cooperative efforts of experts at the three principal manufacturers, Alco, Baldwin, and Lima, working closely with a well-trained and experienced PRR staff, led to a model that offered real promise. Participants in the *S1* project sought to overcome a major weakness of large, fast

During the final days of steam in North America, a *Big Boy*, the largest steam locomotive ever built, powers a long and heavy westbound Union Pacific freight train over Sherman Hill in Wyoming. © William W. Kratville.

locomotives, namely the huge, dynamic forces produced by pistons and side rods that pounded and hence damaged the track structure. Even the most advanced methods of balancing the drivers failed to manage the problem. But designers discovered that by splitting the drive gears into four coupled engines, smaller and hence less destructive power cylinders and gears could be used.

The *S1* was awesome. Its 300-pound pressure and 132-square-foot fire grates allowed this PRR locomotive to generate about 6,000 horsepower and to pull a 1,200-ton train at 100 miles per hour. The *S1* had real "boiler power": the enormous supply of high-pressure steam allowed it to develop more muscle than any other external combustion engine before or after, and no one knew its ultimate capabilities. But as a historian of American steam locomotives, J. Parker Lamb, observed, "The S1, after receiving much national publicity at the New York World's Fair [1939–1940], proved to be both fast and powerful but, in many ways, was 'too big' for everyday work and extremely slippery" (Lamb 2003, 156). After the war, the *S1*s, which PRR employees called simply the "Big Engine," faded from service, and subsequently these 6-4-4-6s became heaps of scrap metal.

As World War II drew to a close, the PRR continued to experiment with the *S1* concept. The result was the *S2*, a duplex locomotive that featured a 4-4-6-4 wheel configuration and could generate nearly 8,000 horsepower at a speed of approximately 60 miles per hour. The prototype performed well, convincing motive power personnel to proclaim the *S2* to be "the most powerful steam locomotive in the higher speed range" (Lamb 2003, 156). This positive assessment led the PRR to build twenty-five more *S2*s at its sprawling Altoona, Pennsylvania, shops. Assigned to fast-freight service, the statistical data for this new PRR product was understandably impressive: 300 pounds per square inch of boiler pressure, a total engine weight in excess of 600,000 pounds, and a tender capacity of 40 tons of coal and more than 19,000 gallons of water. Yet these twilight creations of the PRR steam makers were hardly perfect machines; in fact, the *S2*s would be overtaken by diesels before Altoona ironed out the developmental design flaws.

The PRR also became widely recognized for its *T1*-class nonarticulated passenger locomotives. Early in the war years, these monster steamers with their 4-4-4-4-type wheel arrangement appeared on the property and were capable of pulling eleven 8-ton cars at 100 miles per hour. What often caught the eye of trackside watchers was that the shop workers had encased the engine with a metal shroud conceived by the world-renowned industrial designer Raymond Loewy. They may have also noted that the sixteen-wheel streamlined tender was enormous. When fully loaded with coal and water, it weighed about 80 percent as much as the locomotive itself.

It is significant that the PRR personnel who set the course on steam power may have committed a serious mistake in their overall game plan. In the 1930s engineers and technicians at the major commercial manufacturers had made substantial progress with the development of Northern- (4-8-4) and Hudson- (4-6-4) type engines. In both cases, improvements led to faster and more powerful machines. Better combustion chambers, more efficient value gears, and longer wearing bearings were some of the important improvements. Rather than adding Northerns and Hudsons to its stable of iron horses for freight and passenger operations, the PRR decided to *skip* the 4-8-4 and 4-6-4 experience and move directly to the large duplex stage. "These engines were much more complex pieces of machinery than conventional locomotives," concluded historian Lamb. "As with any new engine configuration that is designed to cure one set of problems, the duplex also created another set of new challenges" (Lamb 2003, 160). Alas, the *Q*s and *T*s had relatively short careers. In hindsight, the PRR could have better spent its $25 million commitment to this motive power had it invested in either the tough and durable Northerns and Hudsons or accepted the coming diesel-electric revolution.

In the immediate post–World War II era, not everyone would readily agree that a particular company should dieselize immediately. Take the case of the Norfolk & Western Railway (N&W). This principally coal-hauling road remained highly profitable with steam power. Its well-maintained fleets of 4-8-4s, 2-6-6-4s, and 2-8-8-2s effectively hauled freight and passenger tonnage over the hills and dales between the Ohio River and the Atlantic Ocean. The 4-8-4s (or *J*s) became legendary for at times pulling passenger trains at sustained speeds of 70 miles per hour. Indeed, these popular engines embodied everything that N&W personnel believed should be part of the modern steam locomotive: one-piece cast-steel bed frame, large boiler, mechanical and pressure lubrication, valve gears, and roller-bearing axles. "The 4-8-4 really has more performance built into her than the short hauls of N&W will permit," opined David P. Morgan Jr., editor of *Trains* magazine, in the early 1950s. "The longest possible haul is the 676½-mile run between Cincinnati and Norfolk, but the schedule cycle . . . seldom allows even that luxury. As it is, the 14 engines in the fleet roll up 500 miles a day each, 15,000 a month—which is commonly regarded as good *diesel* passenger mileage on most roads." Furthermore, there existed a strong tradition of steam locomotive excellence on the N&W. Company craftsmen were widely recognized for their talents for design and construction expertise. "Nobody ever taught the N&W much about steam," remarked Morgan. "They wrote the book" (Morgan 1954, 25). In the postwar years an arguably obsolete technology was not without merit; the N&W offered ample proof. And in other nations, for example China, India, and South Africa, modern steam locomotives capably met the power needs of widely dissimilar rail operations.

TURBINE LOCOMOTIVES

At times technology replacement may be driven by powerful *external* forces. During steam's last years, one example stands out. Several large Eastern bituminous carriers had no desire to offend the domestic coal industry; income from haulage of black diamonds represented a substantial amount of their freight revenues. The Chesapeake & Ohio (C&O), N&W, and PRR (which had stock control of N&W) seemed not about to give up on improving steam power technology. These carriers turned to what became the last commercially practical option: a coal-burning, nonreciprocating steam-turbine locomotive.

The concept of applying steam-turbine technology to railroad locomotives did not represent lunatic-fringe thinking. By the era of World War II, this energy source had been successfully used in ships and for stationary power plants. Although in 1938 General Electric built a high-speed, two-unit, 5,000-horsepower, steam-turbine-electric prototype for the UP, the PRR was the first of the big three coal carriers to try the steam turbine alternative. In 1944 the company took delivery from Baldwin and Westinghouse of a costly 580,000-pound engine that featured two impulse turbines (a distant relative of the water wheel): a 6,500-horsepower main turbine for forward movement and a smaller 1,500-horsepower turbine for reversing. A conventional boiler fed the turbine that was geared to the eight driving wheels. Even though this experimental locomotive performed reasonably well, within a few years the costs of excessive maintenance promoted its retirement.

In 1947 the C&O also turned to Baldwin and Westinghouse for three steam turbine locomotives. These long, heavy units contained a DC generator that drove traction motors, just as with diesel-electric locomotives. Instead of an impulse turbine used by the PRR counterpart, the C&O units featured a rotary machine of the so-called reaction type. But mechanical kinks and other problems led C&O management wisely to turn to the ever-better and equally powerful diesels.

In the early 1950s N&W received from Baldwin-Lima-Hamilton, successor company to Baldwin and Lima, a coal-fired steam turbine locomotive. More closely resembling the C&O units than the PRR locomotive, this final representative of a small, short-lived breed also used a reaction turbine to power DC generators. But its overall mechanical features held important differences from earlier models; for one thing, it contained a specially fabricated watertube firebox, a product of the engineering skills of master boilermaker Babcock & Wilcox. N&W unit No. 2300, officially *TE1* (which stood for turbine electric), was popularly known as *Jawn Henry*, inspired by the railroad folk hero John Henry. It impressed many, including railroad writer Morgan. As he commented in the November 1954 issue of *Trains*, "I can report that *Jawn* rides well, displays a phenomenal ability to make steam, and pulls the dynamometer car needles around with gusto. Once while starting (with the speedometer reading 1 mile per hour) the gauge showed 224,000 pounds tractive effort" (Morgan 1954, 30). Inside its expansive body was a coal bunker, cab, 600 psi watertube boiler, steam turbine, DC generator, and auxiliaries. No. 2300 rode on four three-axle trucks fitted with a dozen traction motors. Even though *Jawn Henry* had admirers, the N&W did not acquire any others. Its high initial costs, maintenance considerations, and enormous length (106.5 feet) killed off this coal-burning alternative to contemporary diesel-electrics. In 1957 the lone N&W steam turbine, which had settled into pusher duty east of Roanoke, Virginia, ended its revenue-producing career.

The first cousin of the steam-powered turbine locomotive was the gas-turbine. The UP spearheaded this new approach to motive power. In 1949 the company took delivery from General Electric of a 4,500-horsepower, gas-turbine prototype and immediately launched extensive testing. The results impressed many at UP, including its president, who observed, "This is jet propulsion on wheels. It might well revolutionize American railroading" (quoted in Klein 1989, 497). Subsequently, the company ordered another fifteen. Generally, performance remained good; these gas turbines provided payload pulling muscle at an economical cost. But problems developed with the residual oil burned by the turbines, and so mechanical engineers decided that propane would be a better fuel choice.

This interest in turbines continued at UP. By the mid-1950s, thirty 8,500-horsepower units had arrived and the company proudly claimed that these iron horses were the most powerful locomotives in the world. For the UP they were the *Big Boys* of the poststeam era. Yet a decade later management concluded that its fleet of turbines should be retired, largely because of unacceptably high maintenance costs. In 1969 the last turbine locomotive pulled its final revenue train and found its way to a public park in Omaha, Nebraska.

About the time coal and gas turbines made their debut on American railroads, a radical motive power concept received a modicum of attention. Surprisingly enough, engineers, scientists, and others toyed with the notion of an *atomic*-powered locomotive. What attracted a serious examination of this application of the mighty atom was the campaign waged in the early 1950s by President Dwight D. Eisenhower and his administration to develop peaceful uses of atomic energy. The Cold War between the Soviet Union and the West continued after the Korean Conflict, and U.S. officials wished to explore every practical aspect of the dawning Atomic Age. As the railroad industry publicly commented, "Of all forms of land transportation, railroads offer the greatest opportunities for the efficient use of nuclear energy. From a domestic perspective surely an attractive feature of harnessing the atom was the feeling that it would be an incredibly *inexpensive* source of energy—power literally "too cheap to meter" (Waite 1996, 37).

When it came to transportation, atomic energy seemed to offer much. Following World War II, the U.S. Navy saw the possibilities for atomic-powered submarines as a practical way to eliminate the need for frequent surfacing in order to refuel. Indeed, in the mid-1950s a technological triumph occurred; on January 17, 1955, the atomic submarine USS *Nautilus* entered the Navy's fleet. Meanwhile, developmental work started on an atomic airplane; this would be a transportation revolution, indeed.

Although the railroad industry thought it wiser to invest its capital into replacing steam with diesel-electric motive power, outside interests began preliminary work on applying the atom to land transportation. Spearheading this practical possibilities of an atomic-powered locomotive was an experienced atomic scientist, Lyle Borst, who in the 1950s was a member of the physics department at the University of Utah. Early in 1954 Borst captured considerable national, even international, attention when he announced that "an atomic powered locomotive has been found to be technically feasible." He also noted that "problems of health and public welfare are considered technically soluble," although he admitted that "the exact arrangements have not been detailed" (quoted in Waite 1996, 39).

What did Borst and his associates have in mind for what they called the *X-12* locomotive? The basic (and hence most discussed) component was the nuclear reactor, which would be provided by boilermaker Babcock & Wilcox. Consisting of two units, the locomotive would have the reactor, "conventional" turbine, and associated traction equipment in the 100-foot-long lead unit. The trailing unit, which measured about 60 feet, would house the radiators for the cooling water. The total weight of the *X-12* would exceed 700,000 pounds, largely because of the immense quantity of radioactive shielding material needed to cover the reactor.

If built, *X-12* would have plenty of pep. Its normal horsepower rating would be 7,000, but it would also have an overload capacity of 10,000 to

12,000 horsepower. Moreover, the *X-12* would have the capability of accelerating a 5,000-ton train from 0 to 60 miles per hour in only 3 minutes, 32 seconds. What this meant was that the atomic-powered locomotive conceived by Borst would possess the strength of a diesel and the pickup of an electric. It would be the ultimate hybrid locomotive.

Advocates of turning to atomic motive power repeatedly emphasized the cheapness of fuel. They admitted that the initial price for the unit would be more than $1 million (twice or three times that of a comparable diesel), but on a ten-year amortization of costs, a diesel would be more expensive, largely based on the fuel differential. Yet, such an optimistic expectation was not grounded in hard numbers. Because of national security, the Atomic Energy Commission classified the price of uranium (U-235), and so Borst and others could only speculate on fuel costs. Since they anticipated some onboard fuel-reprocessing equipment, they really believed that their locomotive could go years without refueling.

Still, the atomic-powered locomotive remained only an idea. Although there would be numerous technical and engineering hurdles to overcome, the one major drawback of any atomic conveyance was the possibility of a serious accident. Since the *X-12* would be assigned to either fast-passenger or fast-freight service, chances for a derailment increased, potentially causing a grave human and environmental emergency. (Of course, the reactor vessel would be protected and energy cooling equipment would be included.) Furthermore, there existed real concern about the overall costs of construction and operation. And if turbines were eventually to prove workable with coal, that seemed to many observers a better option for improving and even truly revolutionizing railroad motive power.

While discussions centered on atomic-powered locomotives, the Atchison, Topeka & Santa Fe (Santa Fe) toyed with the idea of using an atomic bomb to shorten its route across the Mojave Desert in California. Santa Fe engineers took seriously President Eisenhower's "Atoms for Peace" program "to harness the atom for the benefit of mankind," considering what they delicately called "the nuclear option." The roadblock for the railroad involved the Bristol Mountains that rose sharply and suddenly about 1,200 feet from the desert floor. Their plan, which fortunately did not move beyond the "talk stage," involved placing about a score of atomic bombs beneath the surface of these barrier mountains and then vaporizing them. Similarly, highway engineers, who were in the process of building Interstate 40 in the general area, considered the bomb idea but, like their railroad peers, wisely opted for traditional methods of construction.

DIESEL LOCOMOTIVES

Then came the diesel. As World War II ended, hundreds of diesel-electric switch engines daily shunted thousands of cars throughout the United States, doing so with efficiency and cost effectiveness. And an increasing number of snappy, stainless steel, diesel-electric-powered streamliners "highballed" fast "name-trains" between major urban centers. In the Midwest, for example, *Zephyrs* became a synonym for first-class service in scores of cities served by the Chicago, Burlington & Quincy. But the locomotive that really brought enormous change to the railway world was the diesel-electric freight unit. The case can be made that the *FT* (freight), the first widely used diesel freight locomotive, introduced in 1939 by the Electro-Motive Division (EMD) of General Motors (GM), might have been the most influential piece of motive power since George Stephenson's *Rocket* more than a century earlier. In one stroke the *FT* broke steam's historic monopoly of freight traffic and thereby forecast nearly total dieselization, both in the United States and abroad (except where heavy-duty electric locomotives performed over-the-road freight chores).

The EMD *FT* rapidly became the locomotive of choice for many carriers. Shortly after the first set of units entered revenue service on the Santa Fe in early 1941, informed railroaders knew that major technological change was at hand. The possibilities for "electrification without wires" seemed enormously promising. Motive power experts really liked everything about *FT*s. They were impressed with these diesel's durability and dependability, made possible by each unit's sixteen-cyclinder, two-cycle, V-type diesel engine. (It was this style of engine that in 1934 had powered the *Pioneer Zephyr*.) By combining cab and booster units, it was possibly to multiply pulling strength from 1,300 to 2,700, 4,050, and then 5,400 horsepower. No wonder the more innovative railroads clamored for these EMD products. Their stellar performance convinced the all-powerful War Production Board to allow EMD to produce between 1941 and 1945 more than 1,000 *FT* units.

With return to a peacetime economy, EMD lead the way in diesel production. Although other competitors existed, including Alco, Baldwin, Lima-Hamilton, and Fairbanks-Morse, the General Motors subsidiary became the dominant manufacturer. Of course, EMD continually improved its product lines. The immediate successor to the immensely successful *FT* was the *F3*, and thus the "first-generation" diesel-electrics had been born.

As with the *FT*, EMD encouraged railroads, who wished to see how diesels performed on their lines, to try out demonstration units. What took place on the Missouri-Kansas-Texas Railroad (Katy) in 1946 is typical of what North American carriers did and what they decided about this technological marvel. For nearly ten days, officials, who wanted to determine "to what extent diesel power might be successfully used in handling tonnage freight with required speed," paired the three-unit, 4,500-horsepower demonstrator (A-B-A, with the A's having cabs) with the road's finest 2-8-2 (Mikado) steamers on assignments between St. Louis, Kansas City, and Houston. The sixteen-cylinder diesel engines, with their DC generators, AC alternators, and traction motors, performed as intended and trotted tonnage at speeds that at times reached 65 miles per hour. As with *FT*s, all diesel engines in the *F3*s of two or more units were controlled simultaneously; each notch made by the engineer on his throttle changed speed accordingly. The EMD equipment impressed even the most optimistic diesel supporter, especially when the units started a 4,000-ton freight train from a dead stop on a 1.5 percent grade on the main line in Oklahoma after an air hose had snapped. Traditionalists, who held deep admiration for the Katy's 2-8-2s, admitted that this steam power could not have performed such a feat. Some officials then and there decided that, as one person put it, "the diesel is the answer to the problems of tonnage and speed." It would not be long before the Katy ordered from EMD seven A-B-A *F3* locomotives (Hofsommer 1977, 130).

The diesel offered more than "tonnage and speed." For one thing, there were considerable fuel economies. The diesels consumed nearly 8,700 gallons of fuel in tests over the Oklahoma trackage, while the eight oil-fired 2-8-2s had burned nearly 24,000 gallons. In terms of cost, this difference represented more than a 35 percent savings in favor on internal combustion, translating into a cost savings of 8 percent. And there was more. The visibility for engineer, fireman, and head brakeman in these EMD locomotives was far better than on any steam locomotive, allowing for increased safety of operations. Moreover, interior space was attractive and comfortable for crew members. The Katy's employees' magazine described the cabs as having "appointments and conveniences designed to make them only slightly less comfortable than a living room" (quoted in Hofsommer 1977, 130). Employees, too, liked that they did not have to "turn" the diesels on turntables or "Y" track configurations; engineers and firemen could use either the front or rear cab. With dieselization, it did not take long for the venerable turntable to largely disappear from terminals and stub lines.

When EMD pioneered its *F*s, it also introduced the dynamic brake into general use. Although the concept was not new—dynamic brakes had been used on electric locomotives—the EMD product effectively controlled a freight train on its downhill descent by using the traction motors as generators to retard instead of propelling the driving axles. Moreover, this braking system greatly reduced wear of car wheels and brake shoes. The success of dynamic brakes further helped EMD to market its diesel-electric units both domestically and abroad.

Once roads tested diesel-electric motive power, they usually placed their orders. Fast-talking diesel salesman did not face much consumer resistance. After all, the heavy wartime freight and passenger traffic had literally worn out thousands of steam locomotives. "By the end of the war [1945] most of these old engines were ready for the junk heap," remembered a shopman for the Erie Railroad. "It was time for diesels." In their efforts to dieselize quickly, some railroads bought units from multiple manufacturers. The Chicago, Rock Island & Pacific Railroad (Rock Island), for one, practiced what might best be described as the "Noah's Ark" philosophy of diesel buying. The company bought "two of each." While an exaggeration, the Rock Island wished to dieselize as rapidly as possible and so acquired its new motive power from all manufacturers, including the smallest. This practice required the Rock Island to maintain a substantial inventory of parts, but "the superiority of the diesel to the steamer was worth the extra expense and inconvenience."

These first-generation diesel-electric locomotives were constantly featuring notable technological improvements. Although EMD quickly emerged as the dominant maker, exploiting the resources of its direct connection with

When the Erie Railroad celebrated the centennial of the opening of its trans–New York main line between "the ocean and the lake" in 1951, company officials decided to send on tour a highly polished American-type steam locomotive of the nineteenth century and a freshly scrubbed F-type diesel-electric locomotive of the immediate post–World War II era. Courtesy of the author.

An Erie company photographer captured the interior of a F unit. The diesel engines extend along the left side of the picture. Courtesy of the author.

automaking technology at General Motors, less well-established diesel producers also made significant contributions. Take the example of engine cooling systems. About 1950 the Lima-Hamilton Corporation (LHC), which specialized in heavy-duty switchers, did much to perfect this vital part of the diesel locomotive. Company engineers developed a unique water-cooled exhaust manifold that featured cylinder liners cooled for the entire length of piston travel (leaving no hot spots) and forged, heat-treated aluminum pistons cooled by a lubricating oil that passed up from the crankcase. And these innovators at LHC conceived of the intercooler, the device chilling the intake air, that was novel in locomotive application and an important addition to the cooling system.

Railroad managers liked more than what diesel locomotives were doing for the bottom line. They saw this replacement motive power as truly representing modernity. Official after official wanted to rid their properties of the artifacts of Age of Steam and to reveal to the world that railroads were not fossilized giants of the misty past. In 1950 management of the long-bankrupt Georgia & Florida Railroad (G&F) arranged for delivery of several diesel units for its switching and over-the-road freight operations. Immediately, the company placed them on public display, an event that local newspapers and radio stations enthusiastically covered, announcing that "these marvelous machines" would surely "save" the "God Forgotten." Diesels bought time and ultimately the sale of the G&F to the Southern Railway, and a happy ending to decades of court-controlled receivership. Railroads also took great pleasure in showing off to investors, shippers, and the public their bright, new, immaculate diesel shops. These facilities were radically different from those of the old roundhouses and backshops; the nature of the diesel equipment demanded a degree of cleanliness unheard of in the steam era.

The carriers also applauded the quest for technology perfection that diesel locomotive manufacturers sought. By the 1950s, what might be considered "second-generation" diesels began to the appear. An important part of this genre was the "Geep." In 1949 EMD introduced the *GP* (for "general purpose") type of diesel motive power, and this locomotive became an enormous hit. By the mid-1960s the company had produced more 9,000 units (as compared to about 7,400 *F*-type cabs in all series). The strategy behind Geep development was colorfully explained by its principal designer, Richard Dilworth: "In planning the GP I had two dreams. The first was to make a locomotive so ugly in appearance that no railroad would want it on the main line or anywhere near headquarters, but would want it out as far as possible in the back country, where it could really do useful work." As Dilworth continued, "My second dream was to make it so simple in construction and so devoid of Christmas-tree ornaments and other whimsey that the price would be materially below our standard main line freight locomotives" (quoted in Reck 1954, 96).

While the "ugly" objective was tongue-in-cheek, Dilworth and EMD created an affordable, all-duty piece of motive power: bidirectional, accessible, and functional. And there was the "personal touch," for Dilworth had another bright idea. "Before settling on the arrangement of the cab, he built a wooden mockup, windows and all, and invited engineers from the railroads to sit in it and tell him where to put controls, seats, armrests, and other devices that affected the operator" (Reck 1954, 97). The results were gratifying for everyone involved with the *GP*s.

Soon the "age of the *GP*s" had fully arrived. By the early 1950s these workhorses, *GP7*s and *GP9*s, appeared on main lines, branch lines, and shortlines; pulled both passenger and freight trains; and served as worthy prototypes for a series of Geeps, including the *GP20*, *GP28*, *GP35*, and *GP40*. In time these money-making products of EMD engineering featured dynamic braking, turbo-charged V-16 diesel engines, and enhanced horsepower. And the few early production models that lacked roller-bearing wheels were subsequently upgraded, and no more *GP*s were sold without them.

Executives at EMD and at other American diesel makers did not care for a European variation on the diesel-electric locomotive. By the early 1960s the German locomotive builder Krauss-Maffei (KM) began selling to American customers, including the Denver & Rio Grande Western (D&RGW) and Southern Pacific, high-horsepower diesel-*hydraulic* motive power. Unlike a diesel-electric in which power is developed by one or more diesel engines and then converted to electrical energy and delivered to axle-mounted traction motors for propulsion, the diesel-hydraulic, which also has a diesel plant, sends power through a hydraulic transmission by means of shafts and gears to the driving wheels. Supposedly the greatest advantages of the KM diesel-hydraulic involved the production of greater adhesion to the rails, the avoidance of expensive and time-delaying electrical failures, and a superb braking system. Although an employee of the D&RGW described a KM road unit positively by stating, "It's a sweet little engine—really a gem," competition from EMD and other domestic manufacturers together with overall cost considerations limited diesel-hydraulics to mostly European and other non-American venues (Morgan 1961, 48).

Throughout the United States railroad shopmen, who had to be retrained to work on diesels, did not always welcome what became an utterly massive replacement technology experience. They knew, of course, that they had to forget their enormous knowledge of steamers and learn the ins and outs of diesel-locomotive maintenance. After all, on the eve of World War II there were more than 40,000 steam engines and twenty years later only about 200. Still, even though diesels represented a quantum leap in efficiency and economy, some shopmen grumbled about having "those streetcars" (i.e., diesels) in their midst. Workers commonly regarded themselves as craftsmen and did not care to have their considerable skills made obsolete. Where in the past the shops' facilities had built virtually everything for steam power, they now stored parts for diesels. "You become a parts exchanger," noted a Union Pacific mechanical officer. "Very, very seldom do you build anything for a diesel" (Klein 1990, 116). Early on, the joke became that on a steam locomotive it took five minutes to find a problem and five hours to fix it; on a diesel it took five hours to find the problem and five minutes to fix it.

With the diesel motive power revolution, which accelerated dramatically after World War II, railroad managers really did not worry too much about the impact of this replacement technology on their workers. "A new technology does not add or subtract something," explained Neil Postman in his provocative book, *Technopoly*. "It changes everything." And he added, "Those who cultivate competence in the use of a new technology become an elite group that are granted undeserved authority and prestige by those who have no such competence" (Postman 1992, 18). So it became a common occurrence for that old boilermaker not to find an assignment in the diesel shop and the steam locomotive engineer unable to become "qualified" to operate a diesel locomotive. Their old expertise became worthless, and these men suffered accordingly. Many took employment elsewhere or retired.

ROLLING STOCK

Another way that railroad executives made their industry look up-to-date was in operating the latest passenger rolling stock. The time of Old Age also became the time of stainless steel. It would be the Philadelphia-based Budd Company that did much to make this metal practical for lightweight passenger-car construction. The real breakthrough came when the firm perfected a method of electric, controlled-energy welding known as the "shotweld process." The result was that the combination of stainless steel and shotwelding, which produced an attractive and amazingly durable product, revolutionized the car maker's world.

If there was a major negative to Budd's transformation of passenger rolling stock, it was price. Stainless steel was the most expensive material employed for railway car manufacturing. Even during the depths of the Great Depression, the price far exceeded that of conventional steel. But railroads paid top dollar for this equipment, because it often lasted for decades in regularly scheduled service, easily amortizing the investment.

There were additional concerns. Some companies complained that it was difficult to keep stainless cars bright, and they fussed too about the cost and inconvenience of any major repairs that resulted from serious derailments. In the latter case, it might become necessary to return the damaged car body to the Budd production plant for frame straightening and replacement parts. In fact, this was the reason that the Portuguese National Railways in the 1950s decided not to add to its fleet of stainless cars.

Although aluminum and stainless steel worked well for building passenger cars, the all-metal era (including freight equipment) soon relied heavily on Cor-Ten steel. Developed in the mid-1930s by metallurgists at United States Steel, this product of carbon steel together with small quantities of chromium, copper, phosphorous, and silicon made for an inexpensive, strong, and rust-resistant alloy. At first rolling stock that used Cor-Ten steel was usually riveted, but then welding became the preferred mode; perfection of electric welding gear made this an economically attractive alternative.

Yet it would be a mistake to assume that all passenger carriers quickly opted for lightweight streamlined equipment. The largest and wealthiest roads made this investment but only for their important trains. Older, yet still operable heavyweight equipment became assigned to lesser trains. And some roads, whether the Akron, Canton & Youngstown or the Minneapolis, St. Paul & Sault Ste. Marie (Soo Line), never moved beyond either the wooden or heavy-steel eras. Until the demise of privately owned intercity passenger service in the early 1970s, less than a quarter of the total passenger car fleet consisted of lightweights. After all, the overall quality of the existing heavyweight cars was too good to discard, especially considering the substantial capital expenditures that lightweights required. For some roads it made more sense to modernize their heavyweight rolling stock, usually with new seats, lighting, and floor coverings. It was not that the new technology was unpopular, but rather that it was generally too costly to employ widely.

By the 1940s passengers who rode on the principal trains of American railroads no longer needed to worry about soiled or burnt clothing during warm weather because of open windows that allowed occasional ash, cinders, and soot to shower them. Dieselization had mostly taken care of that annoyance. In fact, a few railroads in the East, most notably the Delaware, Lackawanna & Western (Lackawanna), had for decades burned hard coal in their passenger locomotives, mostly eliminating the problem. The Lackawanna became widely acclaimed for its popular advertising campaign that

In the late 1940s the St. Louis–based American Car and Foundry Company built a "futuristic" streamliner known as the *Talgo*. But its lightweight construction produced an uncomfortable ride and the train set remained only a failed prototype. Courtesy of the author.

began about 1900 with the mythical Phoebe Snow, a beautiful young woman, always clad in white linen—cool, comfortable, and unruffled. The public loved that Miss Snow spoke only in rhyme, the most famous being: "Says Phoebe Snow/About to go/Upon a trip/To Buffalo: 'My Gown stays white/From morn till night/Upon the road of anthracite.' "

Yet if a steam locomotive, burning soft coal, pulled the main line train, passengers likely avoided the smoky residue because their cars were equipped with air conditioning. Although the origins of air conditioning in railway cars dates back to a U.S. patent granted in the mid-1850s for an ice system with axle-drive fans, development was slow. The industry had other concerns, and, of course, commercial structures, whether hotels, restaurants, or theaters, at best had only fans (kerosene prior to electricity). Before World War I, however, the Santa Fe gained some acclaim for its electric fan, ice-cooled ventilating system in its dining cars that were part of Los Angeles–bound luxury trains. But this primitive air-conditioning arrangement worked poorly, and within a decade or so it was discarded.

It was the Baltimore & Ohio Railroad (B&O), and not the Santa Fe, that led the way in modern railway car air conditioning. In the late 1920s the B&O hired Willis Carrier, a pioneer of commercial air conditioning, to keep a passenger coach and then a diner cooled. His approach featured an electric motor attached to an ammonia refrigeration compressor. The latter unit cooled water that was pumped to coils located in the roof ducts of the car. But difficulties developed, and so the B&O turned to the York Ice Machine Company for what turned out to be a better approach, namely, cooling units powered by gasoline engines. Specifically, this equipment circulated a brine coolant through ceiling coils. Not long afterward it would be the indefatigable Carrier who devised a special coolant that replaced the brine water. The investment that the B&O made in outfitting its first-class passenger equipment drew considerable attention. An editorial writer for the *Los Angeles Times*, for one, commented that a trip on the air-conditioned B&O *Columbian* made for a pleasant summertime experience, and complained that in the West "we refrigerate our fruit but roast our passengers" (Stover 1987, 288).

Except for the Santa Fe, B&O, Pullman Company, and a few other carriers, railroads did not show much enthusiasm for cooling their passenger rolling stock. Factors of cost, service location, and tradition did not usher in quickly the age of air conditioning. By 1940 only about 12,000 of the nearly 60,000 cars in service in the United States contained an ever-improving system of air conditioning, although by 1950, the total fleet of air-conditioned equipment had increased by 5,000. Yet by this time, passengers who traveled first class, including all Pullman patrons, almost always enjoyed comfortable, climate-controlled chair and sleeping cars, and those carriers, especially in the South and West, that could offer this modern technology made much of it in their advertisements.

Air-conditioned comfort was a standard feature in a dramatic improvement in the self-propelled railway passenger car. This would be the all-stainless steel rail diesel car (RDC), which the Budd Company introduced in 1949. It did not take long for this updated version of the classic "doodlebug" and the several short "streamlined" motor trains of the 1930s to appear throughout the world, being especially popular in North America. Although companies commonly assigned these 59-ton RDCs to secondary, branch line, and suburban runs, the New York Central, for one, used its first two RDCs in high-speed service between Boston and Springfield, Massachusetts, a distance of nearly 100 miles, and the Western Pacific placed two RDCs, known as *Zephyrettes*, on a grueling 924-mile run between Oakland, California, and Salt Lake City, Utah.

The Budd Company built its RDC fleet with the latest technologies. It selected two ultramodern, 275-horsepower General Motors diesel engine power units that earlier had been designed for rugged duty in army tanks. Each engine, mounted under the floor, drove one axle, providing independent operation and notably increased traction. Power was transmitted by a GM torque converter and reverse gear fully synchronized with the engine. Trucks were equipped with disc brakes, anti-wheel-slide devices,

In efforts to reduce the cost of commuter, local, and branch line operations and to make passenger service more attractive, railroads worldwide showed interest in the rail diesel car (RDC). These self-propelled units became popular because they could be operated separately or lashed together in a trainlike fashion. Courtesy of the author

and automatic sanding units. The results produced a passenger unit that cruised at 70 miles per hour, offered quick starts and stops, operated in multiple units, and ran economically. The success of this equipment encouraged other imitators, although Budd controlled the domestic market. Even though demand for this type of vehicle collapsed as passenger service waned in North America, a few RDCs continued to operate in the twenty-first century, attesting to the engineering wizardry of the now-defunct Budd Company.

A reading of the popular press from the 1920s through the 1950s reveals virtually no mention of the quiet technological revolution that affected the lowly freight car. These ubiquitous pieces of rolling stock not only became larger and were made mostly of steel construction, but also contained vastly improved braking systems. Not long after the start of the twentieth century, a basic change in freight-car brakes came about with introduction of the "Type-K" unit, designed to control the longer trains of seventy-five to 100 cars that were becoming more prevalent on busy main lines. This replacement breaking device provided quicker action because of accelerated air input, making possible stops in about a third of the distance that the then-conventional air brakes allowed.

Even as Type-K brakes emerged as the industry standard, the ICC, the federal regulatory body that had a federal mandate to create safe railways, began in the 1920s to push for an even better braking system for freight trains. In time the ICC concluded that the triple-valve concept could not be perfected beyond the Type-K's. Prodded, therefore, by the regulators, the air-brake industry introduced an entirely new control valve known as the "AB." This replacement system contained two major control valves in a single body, one for regular stopping and one for emergency. The rate of transmission for the AB valves was about 40 percent faster than the Type-K and took advantage of the greatest possible speed of an air wave in a pipe (900 feet per second). Unlike the Type-K's, the brake cylinder and reserve tank were separate units. In 1933 the Association of American Railroads, the industry's trade group, adopted the AB approach, and shortly after the end of World War II more than 60 percent of the American freight fleet featured AB devices. Soon all "interchange" cars, which moved from one railroad to another, had AB brakes. Improvements continued. Introduction of composite brake shoes with their high coefficient of friction, for example, made it easier to stop even trains that exceeded 100 or 150 cars. Yet, braking remained "mechanical" on freight trains, although passenger streamliners customarily had more advanced, electrically controlled air brakes.

PIGGYBACKS

Beginning in the mid-1930s, a new kind of freight-train cargo that caught the eye of trackside watchers was the trailer-on-flatcar or, as the industry came to call it, "TOFC." The public, however, preferred the more picturesque "piggyback" moniker. The replacement technologies of diesels, AB brakes, and the like did much to make TOFC a competitive and profitable transportation service.

Innovations in the railroad industry have repeatedly come from smaller, less entrenched carriers. The reason is understandable. New methods and devices often aided financial survival, or at least that was the intention. This holds true for piggyback transport.

The Chicago Great Western Railroad (CGW), a now defunct or "fallen flag" Midwestern carrier that served the gateways of Chicago, Kansas City, Minneapolis–St. Paul, and Omaha, developed the modern piggyback concept. It became the first *steam* road to operate this service over a relatively long distance and on a permanent basis. In the summer of 1936, the CGW began to haul commercial trailers regularly, continuing the practice without interruption until 1968 when it became part of the Chicago & North Western Railway System.

The notion of moving roadway vehicles by railroad predates the 1930s. For years, steam railroads, for example, transported circus equipment and customarily placed animal-driven wagons and later truck trailers on flatcars.

Electric interurbans, though, blazed the way. In 1926 the Chicago, North Shore & Milwaukee Railroad (North Shore) started to transport truck trailers, being forced to compensate for its inability to bring carload freight into the heart of Chicago over the local elevated commuter rail system because of its sharp curvature and congestion. Soon several other traction lines in the Midwest followed suit. In 1931 the Chicago, South Shore & South Bend Railroad (South Shore), for instance, announced a "Ferry Truck Service." It advertised in its public timetable, "Trailers left at your door for loading, shipped overnight on special flat cars."

Early in 1935 the alert Chicago-based traffic department of the CGW called management's attention to the piggyback concept as another way of generating badly-needed income. The North Shore's success sparked the suggestion. The leadership believed that it was worth exploring and Samuel Golden, an imaginative official, took charge. A former executive with the Standard Car Company, Golden had already demonstrated a creative bent at the CGW. For example, he had instigated, albeit rather unsuccessfully, what may have been the country's first attempt at aerial weed spraying of rights-of-way. Golden even tried hydraulically controlled highway traffic barriers, but these contraptions did not prove satisfactory. Yet another of his projects, the installation of continuously welded rail (see later in this chapter), fared much better.

Golden pushed ahead, working closely with the CGW's general car inspector in Oelwein, Iowa, who had supervised the movement of the Ringling Brothers and Barnum & Bailey Circus trains over the road. Using circus transport as a model, Golden borrowed a truck trailer from a commercial motor carrier, and shopmen placed the trailer on a standard 40-foot flatcar and attached it as securely as they could. No problems developed with several experimental runs. These tests intrigued the trucking company, which told the CGW that it would move trailers on a regular basis between Chicago and Dubuque, Iowa, if the railroad could provide equitable rates, several specially equipped cars, and necessary ramp facilities.

With this favorable turn of affairs, CGW personnel got ready for commercial piggyback service. Workers equipped ten flatcars: the tiedowns were mostly fashioned from available materials, and a speciality equipment manufacturer provided moveable jacks for trailer support. Although the first customer subsequently decided not to continue with the arrangement, other motor carriers contracted with the railroad. On July 7, 1936, Golden's dream became reality. And all went well, even on the maiden trips between Chicago and St. Paul, Minnesota. Forty-five years later, the general car inspector remembered,

> I rode [from St. Paul] on the flat cars, around and under the trailers, and had instructed the engineer to limit speed to 25 mph. At Empire [Minnesota, Milepost 22], I told him to increase speed to 35 mph to Randolph [Minnesota, Milepost 32]. At Randolph I instructed [him] to resume timecard speed [50 mph] and returned to the caboose, where I rode the balance of the way, but I did inspect the train at every stop. We went through to Chicago without trouble. But I was a casualty, as I got a cinder in my eye going up the Mississippi Hill out of St. Paul and didn't get it out until I reached Chicago. (Quoted in Grant 1986, 33)

The CGW found that its TOFC generally performed smoothly. There existed a few cases where jacks tipped because of improper securement, but employees designed a double jack that corrected the problem. Proof that TOFC equipment needed no major refinement occurred in the winter of 1936 when a trailer train and a passenger train collided at high speed. The wreck left five dead, another six injured, and two locomotives demolished, but "not one trailer was dislodged or broken loose from the car, all were upright." Furthermore, the flatcars, "which were all steel with fishbelly-type side and center sills, bent down in the center and rested on the roadbed, but trailers stayed in place although a few had the ends broken out due to shifted loads" (Grant 1986, 33).

Even though the CGW pointed the way with TOFC for steam railroads, the industry expressed little enthusiasm. There were several reasons for this hesitancy. The technology had not been fully developed, and a general lack of standardization existed among those companies that offered piggyback service. Since the existing offerings nearly always involved loading only a single trailer on a flatcar, profits were modest. Moreover, some officials feared that truck trailers would merely diminish boxcar loadings. Still others, bound by tradition, considered the concept to be only a "flash in the pan" and were reluctant to make financial commitments.

But in the early 1950s resistance to TOFC crumbled. Innovators—most of all Gene Ryan, who launched the Rail-Trailer Company in 1952—responded effectively to the alleged or real disadvantages of the piggyback concept. Ryan and others made such improvements as placing two trailers on each lengthened flatcar (often 75 feet), employing double instead of single jacks, and streamlining loading and unloading facilities. Within a few years, scores of railroads offered customers this service. Railroaders seemed pleased with what was taking place. "This new branch of railroading since its inauguration in July of 1954 by the Wabash has doubled and redoubled," told the Wabash president to a St. Louis journalist in 1956 (quoted in Grant 2004, 216). The company was delighted that it was regaining traffic that had been lost to over-the-road truckers. The Intermodal Age was coming into its own, and as such a parade of new support inventions, including the overhead gantry crane, increased efficiencies and profits.

Although smart intermodal services, with their high cost-to-benefit ratios, became widely discussed during the 1950s, a much improved freight car, introduced as the decade ended, soon became a shining star. This was the 100-ton-capacity covered hopper car, known as the "Big John," designed for handling grain. The determined efforts of Dennis W. Brosnan, creative president of the Southern Railway and a college-trained engineer, made possible this highly productive and popular piece of freight equipment.

The Big John revolutionized grain movements. Before its introduction this commodity commonly traveled in 40-foot boxcars that each carried about 25 tons. The conventional way to move grain by rail was both awkward to load and to unload and frequently leaked along the way, much to the delight of birds and rodents. But Brosnan's new all-metal car, built with the aid of engineers from the Reynolds Aluminum Company, was really a big bin on flanged wheels. It was high capacity, was easy to load from the top and unload from the bottom, and showed the way for the development of other "jumbo" freight cars. Although Big Johns were not prohibitively expensive, their weight required a good track structure with sound rail joints and strong bridges. As a result some branch lines, shortlines, and industrial spurs could not safely accommodate the Big Johns. But if railroads wanted to exploit the advantages of this replacement rolling stock, they needed to spend on track and bridge improvements to make money.

Surely a "first cousin" to the Southern Railway's Big John was the center-flow hopper car that eventually dominated freight equipment in North America. By the late 1930s, covered hopper cars had begun to appear in several shapes and sizes to handle specialized cargoes of different densities, including carbon black, cement, and phosphate. Although this equipment was far superior to long-used open gondola cars (flatcars with short, horizontal wooden walls), these early covered hoppers, with their internal sectional arrangements, were not that easy to unload. Significant amounts of the shipped material would "hang up" around the center sill in these two-, three-, or four-compartment units and remain there.

But this was about to change. In 1961 the American Car & Foundry (ACF) created a demonstrator covered center-flow hopper that would soon fundamentally change this type of rolling stock. What ACF did was to manufacture a car body that was essentially a welded-steel tube that used its sides and side sills to carry the load and also to buffer energy caused by the force of coupling. For the first time the interior of a covered hopper was smooth, with no protrusions to gather contents. A simple ladder allowed a person to climb into the interior for inspection, cleaning, and occasional maintenance.

Some modifications to the center-flow hoppers followed, including increased capacity and slightly curved car-body sides to create what became the trademark "teardrop" shape. In the mid-1960s an imaginative improvement came with introduction of pressure-differential cars (*PD*s), where air or other gases could be added to the car at either the loading or unloading site to facilitate the emptying process. A few years later ACF introduced what it called the conditiontrol car, a covered hopper that was pressure tight and insulated. Shippers now could safely send certain fruits and vegetables in bulk with this equipment. Subsequently ACF cleverly created its Conditionaire, a center-flow hopper with a diesel-powered climate-control system that could keep bulk produce like grapefruits, oranges, and onions from spoiling in hot and humid weather and freezing during the winter.

During the time that Big Johns and center-flow hoppers made their debut, "rack cars" for transporting new automobiles and light trucks entered service. This would be another example of freight equipment becoming more specific to the transported product. When automobiles were first manufactured, they traveled to their destinations in boxcars, the common way to ship most freight. The type of boxcar used, though, was typically 50 feet long and equipped with wide, double doors for easier loading and unloading. But there were drawbacks with this freight rolling stock. Although these "rolling garages," as they were called, offered great protection from the elements, this rolling stock lacked much efficiency—only four vehicles could be placed in a car. Even though these boxcars featured wide, double doors, they were still awkward and slow to unload. Over time the railroads started to lose automobile traffic just as other freight business evaporated, especially following World War II. Better highways enabled trucks to haul all kinds of cargoes long distances, including automobiles and other motor vehicles. Although the automobile boxcar was still employed by the late 1950s, railroads handled only about 10 percent of the automobiles produced.

About this time, railroad managers asked the question: how are we to win back this business? The answer came from Europe. In 1958 the giant German manufacturer Volkswagen shipped its most famous model—the Beetle—by rail to port cities for export. These little vehicles left the factory on a specially designed double-level flatcar that held ten units. Briefly several American railroads experimented with newer equipment, but they were nothing more than an open-sided version of the conventional automobile boxcar. When consumers wanted larger, more powerful cars,

Detroit responded. Soon the 50-foot-long automobile boxcar became obsolete. Although there was some testing in the 1950s of auto transporter truck trailers in piggyback service, it would not until the end of the decade that the American railroad industry decided to copy the German approach of the multilevel flatcar or auto rack. Some of the 85-foot flatcars used for piggyback truck trailers were equipped with rack decking, and hence the auto-rack car was born. Two versions appeared: bilevels, usually holding eight to ten autos, and trilevels that accommodated twelve to fifteen.

As with Big Johns and center-flow hoppers, the early fleets of auto racks breathed new life back into the railroads. Still, this rolling stock required refinements. The principal drawback was that the new vehicles were exposed. Rack cars, most of all, became targets for vandals and thieves; vehicle exteriors were damaged; and parts were stolen. And at times automobiles placed on the open upper deck suffered damage from the exhaust soot if coupled directly behind the diesel locomotives, and from falling icicles from tunnel ceilings and bridges. The solution was relatively simple: rack cars were wrapped with wire mesh and later fiberglass and corrugated metal screening, creating what railroaders called the "paneled" car. These 89-foot auto racks, the newly established standard length, that presently travel on American railroads contain side panels and roofs that fully protect shipments from vandalism, theft, and the elements. Interestingly the transport process has gone full circle, returning to the fully enclosed environment of the original automobile boxcars.

A BETTER ROADWAY

By the 1930s trains that rumbled or even raced over the United States' sprawling network of main lines usually did so over heavier steel rails, crushed-rock ballast, and creosote-treated crossties. Bridges and highway overpasses featured steel and concrete components and handled great weights. After all, by this time the nation's engineering schools had graduated thousands of well-trained civil engineers, many of whom had begun lifetime careers with railroad companies. The railroad engineering departments at the Massachusetts Institute of Technology, Purdue University, the University of Illinois, and the University of Michigan produced some of the best engineers, and carriers eagerly hired them. Even the Great Depression did not prevent railroads from upgrading the quality of their engineering personnel.

It would be during these hard times that one of the more significant advancements took place in the basic design of railroads, namely, the introduction of continuous welded rail (CWR). Just as the creative Chicago Great Western spearheaded modern piggyback operations, so, too, did the company blaze the way for what proved to be a superior alternative to bolted sections of rail. In May 1939 the CGW installed 0.99 miles of this experimental rail on its main line near Oneida, Iowa. Maintenance-of-way employees fused scores of 39-foot sections of 112-pound (per yard) steel rail by using a proven welding technique, the Ox-Weld Automatic Pressure Process that heated the butt ends to 2,280 degrees Fahrenheit. Specifically, six welds took place in the nearby Oelwein terminal yard and they were followed by one field weld. At the staging site employees set up their welding

Beginning in the 1950s more railroads showed interest in continuous welded rail (CWR). Known also as "ribbon rail," these permanently connected steel rails lacked traditional rail joints and therefore eliminated about three-quarters of the costs of track maintenance. Passengers noted that stretches of CWR lacked the rhythmical "clickety-clack" that for generations had been so much a part of rail travel. Courtesy of the author.

machine in a car body and fed rails through the open end doors from an adjoining flatcar, placing them onto a string of twenty-seven flatcars. Then a work train delivered these cars to the installation location.

The Oneida track worked well. Within the first year this track segment endured the pounding of freight trains that traveled an average of 45 miles per hour and passenger trains that ran at 60 miles per hour and consistently maintained its alignment. And the welds generally withstood the extreme temperatures of eastern Iowa, which regularly ranged from 100 degrees in the summer to 10 degrees below zero in the winter. Although seven weld failures occurred during what became a ten-year test period, the CGW was impressed with the results. Company engineers wisely replicated the experiment with similar jointed track. Moreover, the usually cash-strapped railroad liked the decrease in maintenance costs. After a decade of observations, the nonwelded track encountered expenses of $5,075 while the welded track cost $3,190. In time, especially after 1950, other carriers followed suit by installing thousands of miles of CWR, and by the 1980s the practice became virtually universal on main and secondary lines.

COMMUNICATIONS AND COMPUTERS

As the railroad industry entered Old Age, it continued to use well-established means of communications. Well into the 1940s and 1950s, it was still possible to hear the clicking of telegraph keys and sounders in thousands of railroad stations, control towers, and dispatching facilities. Agent-operators and towermen received information about train movements, availability of rolling stock, and various railroad matters through telegraphy. Telephones also remained heavily used and in some cases had replaced the older telegraphic equipment. While improvements took place in both means of communications, especially telephones, it was the two-way radio that became the talk of the industry.

As with dieselization, air conditioning, and other betterments, some railroads exhibited keen interest in replacement communication technologies. One such carrier was the Erie Railroad, a historically financially troubled company whose main line stretched between New York City (Jersey City, New Jersey) and Chicago. The Erie, like some other roads, took considerable pride in its long tradition of meaningful innovations. Early in the twentieth century the Erie pioneered the use of telephones for train dispatching; in the 1920s it led with the "printing telegraph," a forerunner of the teletype, and after World War II it employed some of the earliest commercial "walkie-talkies" for yard operations. Then in 1951, the company

In the early 1950s the old and the new are captured. Although the engineer is at the throttle of a vintage steam locomotive, he is using a telephone, likely talking with a crew member or a company dispatcher. Courtesy of the author.

became the first major railroad to have its entire main line served by radio. Known as the "four-way train radio-telephone," this very-high-frequency (VHF) system allowed static-free voice contact from train cab to caboose, train to train, train to station, and station to station. "The Erie should be congratulated for continuing to set a progressive pace for American railroading," editorialized a Cleveland, Ohio, newspaper in 1950. "[Radios offer] unending possibilities for increased safety and convenience" (quoted in Grant 1994, 52). And this observation proved entirely accurate.

Paralleling the Erie as an innovator, the Chicago Great Western also became an early user of two-way radios. In a clever, innovative way its management introduced this technology to employees, who often resisted change. A company executive told supervisors that they must inform trainmen and on-ground employees what they *could not do* with radios, based on rules established by the Federal Communications Commission. What happened was what management expected, namely, employees found a variety of ways to incorporate radios into their daily work. They appreciated the extensive freedom to act as they saw fit and quickly realized that two-way communications made for more convenient and safer operations. For one thing, in a blinding blizzard or a heavy fog, a brakeman would tell the engineer that a cut of freight cars occupied a siding without having to use the old method of hand signals or a flare, a fusee, or even walking to the locomotive.

Railroad workers mostly understood the value or potential value of modern radio communications. After all, they had home and car radios and were comfortable with the general technology. And in the early 1950s some railroaders had acquired television sets for personal entertainment or at least thought about making this high-ticket consumer purchase. But when it came to computers, they were mostly if not totally in the dark, even those individuals who were involved in office work, whether it be car accounting or disbursements.

The computer age began in the 1950s, and a variety of businesses soon realized the computer's potential usefulness to perform routine tasks. However, this was generally not the case in the railroad industry, even though the nature of the bureaucracy cried out for this technology. By the mid-twentieth century, railroad companies commonly suffered from a certain amount of ossification in their business practices. Old habits took decades to die. Earlier, companies had been painfully slow to adopt carbon paper, preferring instead to remain with the ancient practice of making "wet copies," a process that employed onion-skin paper, moist cloths, and heavy metal presses. Similarly, roads continued to use straight pins, rather than paper clips, to hold together letters, memoranda, and the like. It was not until after World War II that carbon paper and paper clips were at last generally found in railroad offices.

Just as the Union Pacific embraced the earliest passenger streamliner technologies, alert company officials sensed that computers, even in their infancy, might make major contributions. But at the time the road was virtually alone with its progressive thinking. In 1954 UP installed its first computer, supplied by the International Business Machine Company (IBM), at yard offices in three scattered locations. The devices made tracing car locations easier and improved equipment utilization. It did not take long before computers also allowed UP to manage better its interline-freight accounts, material inventory, and payroll. Four years after the UP experiment, the Missouri Pacific Railroad made a modest investment in computers and soon thereafter more roads became believers. In 1962, the Wabash Railroad, for one, acquired IBM 1401 and 1410 data-processing systems, which consisted of computers, punch cards, readers, printers, and tape drives and immediately predicted an annual savings of $65,000 for its accounting operations. In a sense the technology of computers was to the railroads of the late 1950s and 1960s what diesel-powered streamliners had been in the late 1930s and 1940s. Each produced unexpected economies, often beyond the wildest expectations of management. As one official for the Chicago & North Western Railway correctly observed before the company's board of directors early on in the computer revolution, "The potential for computers at my company and in our industry is limited only by our imagination." And he would be correct.

OLD AGE

In its Old Age, the railroad industry felt the sting of competition from automobiles, trucks, and airplanes. Although managements came to comprehend the enormous financial benefits that the diesel revolution produced, by the 1960s they sensed that those economies had run their course. Increasingly industry officials turned to corporate mergers as the principal way to strengthen their bottom lines and, in some cases, to save their companies from insolvency. Therefore, the pressing objective became corporate unification. The widespread "merger madness" of the 1960s and 1970s produced such giants as Penn Central (later Conrail), Burlington Northern, and CSX, which helped to create a new corporate era in the 1980s. It would be during this period that the freight roads shed their money-losing passenger operations, leading to the creation in 1971 of the quasi-public National Railroad Passenger Corporation (Amtrak).

Just as the railroad industry now emphasized *mergers* rather than *technology* for enhanced corporate earnings, it also worked diligently to thresh out problems that came from intense governmental regulation and antiquated labor work rules. On both fronts the industry would score major victories, with the most spectacular being the enactment in 1980 of the Staggers Act that brought about partial industry deregulation of freight rates and as such made the Interstate Commerce Commission superfluous, leading in 1995 to its abolition. No longer was the theme of railroad regulation "enterprise denied," but rather it became "rebirth." A relaxing of rigid and expensive labor agreements likewise contributed to this rebirth phenomenon. Fundamental technological changes, best illustrated by the diesel-electric locomotives with no fires for the firemen to tend, had led to greatly altered employee needs. A give-and-take attitude by management and unions brought about a new and happier day in labor relations.

5

Rebirth, 1970–Present

MOTIVE POWER

By the time the American railroad industry celebrated its newfound freedom of reduced federal involvement in rate-making and operations and began to pare down its large and expensive work force, most notably superfluous locomotive firemen, it remained committed to acquiring the best motive power. It would be during and after the 1980s that "third-generation" diesel-electric locomotives entered revenue service. The force behind these practical pieces of Rebirth locomotion would be the General Electric Company (GE) rather than the manufacturing giant, Electro Motive Division (EMD) of General Motors, which for fifty years had dominated both the domestic and international marketplace.

In the diesel era, railroad officials came to realize that no single acquisition they made was as important or as fraught with peril as locomotives. Unlike steel rails, crossties, or freight cars, diesel-electric locomotives compete on more than cost alone. Builders' products are different: price only matters when everything else is largely equal. In the late 1970s GE reflected on what it had and had not accomplished in the two decades of producing heavy-duty freight motive power. Although the company took some satisfaction in knowing that it had established itself as the alternative to the LaGrange, Illinois–based EMD, it was just that: an alternative. There developed the sense that railroads, both at home and abroad, at times acquired GE products mostly to place price and quality pressures on EMD. When GE introduced its *Dash 8* locomotives, impressive changes occurred.

The *Dash 8* story really began in 1980 when GE engineers initiated a development program to upgrade the *Dash 7*. The goal was to create a highly fuel-efficient big freight locomotive that gave strong tractive effort. Here was a classic case where outside forces had a direct impact on technological advancements, namely, the enormous spike upwards in fuel prices that occurred in the wake of the Arab oil boycott of 1973–1974. Although the *Dash 7*, the premier GE locomotive product of the 1970s, came to achieve up to a 16 percent improvement in fuel efficiency over comparable earlier models, railroads wanted (and needed) more. And, as always, carriers demanded high reliability. So, in the fall of 1982 GE began extensive testing of the prototype *Dash 8*. This 3,600-horsepower locomotive utilized proven components, yet it offered cutting-edge features, including microprocessor control systems. As expected, these newest products from GE were fuel efficient. Even the building process was state of the art. At the GE manufacturing complex, located near Erie, Pennsylvania, robotics, computerized control, and production sequencing became part of the procedure. Quality control stood as the hallmark of this highly regarded GE operation.

Not surprisingly, there would also be a *Dash 9*. In the mid-1990s GE offered customers what it considered to be a locomotive that was bigger and better than ever. And the claim was on the mark. This massive machine, made with 200 tons of steel, measured nearly 74 feet in length, stood almost 16 feet above the rails, and generated a whopping 4,440 horsepower. Improvements, of course, had been made on the *Dash 8* design, with arguably the most important being the inclusion of electronic fuel injection (EFI). Replacing the conventional mechanical fuel injection system, EFI facilitated optimal engine combustion through calculated variation in fuel-injection timing, producing better fuel economy and reducing hazardous emissions. Moreover, EFI allowed for the elimination of some traditional components that were part of mechanical fuel injection, therefore improving reliability and reducing maintenance costs.

It would be in the 1990s that another revolution of sorts affected American-built motive power. This involved the introduction of alternating current (AC) traction for diesel-electric locomotives. In the early 1960s EMD engineers, who had never abandoned their quests for an ever-better product, began constructing and testing prototype AC-traction motors. But not until the late 1980s was there available a compact, durable, and practical switch gear, which was required to control high-torque AC motors at variable speeds. And thus it would be EMD, rather that the new industry leader GE, that in 1993 introduced the 4,000 horsepower *SD 70*. "No locomotive has so completely changed the way railroaders think about the way they run trains since the FT a half-century ago," observed a proud EMD official. "Basically, the SD 70 MAC is a mountain-grade leveler. No one is ever going to seriously consider D.C. for heavy-haul again" (quoted in "10 Diesel Locomotives" 2002, 42). In the best tradition of American capitalism, GE responded; it soon offered customers the improved *Dash 9*. Railroads could choose between the more complex and expensive AC traction for heavy haul and simpler and less costly DC for everything else. (Companies paid about $500,000 per unit more for AC traction.) These third-generation AC and DC locomotives, whether EMD or GE, became particularly popular for heavy-unit coal, grain, and ore trains and for high-speed intermodal service. Crew members, too, liked them. Especially well received were the comfortable and largely soundproofed cabs with their "wide-body" design that provided space for such amenities as high-back adjustable seats, a conductor's desk, and a refrigerator.

INTERMODAL

If there was a symbol for the Rebirth of railroading, it surely would be the intermodal freight train. With the partial deregulation of rates after 1980, carriers could compete much more effectively with long-distance trucking firms. By the start of the twenty-first century, intermodal business has become the leading source of revenue for major North American railroads, surpassing the historic leader, coal. In recent years the old, antagonistic attitude of railroaders toward truckers diminished, and motor carriers have come to view railroads as their transportation "friends." Why this transformation? Intermodal movements helped truckers deal with such nagging matters as driver shortages and turnovers, hours of service, high insurance and fuel costs, diesel emissions, and interstate highway congestion. As with improvements to railroad motive power, the latest technologies were applied to intermodal operations, ranging from loading cranes to new approaches in transporting shipments.

"Containerization" became the trademark of a reenergized railroad enterprise. By the twenty-first century it would be container-on-flatcar (COFC) service that led to the decline of the well-established "piggyback" or trailer-on-flatcar (TOFC) operations. As with so much of the technological advancements in railroading, the concept of containerization predated active usage.

During the era of World War I, when military traffic overtaxed the American transportation system, the modern prototype for freight handing in containers became available commercially. The first phase of containerization took place in Cincinnati, Ohio. And the individual most responsible was Benjamin Franklin Fitch, a onetime employee of the White Motor Company, an early manufacturer of heavy-duty trucks. Fitch designed a highly practical way to transfer less-than-carload (LCL) freight between local freight terminals. By using "demountable bodies" (containers) to gather cargoes at the various locations, a 5-ton White truck could pick up and drop off the body at the delivery sites. Specifically, containers were made from flat, steel straps (later enclosed to protect cargoes), and manual chain hoists allowed workers to manipulate the bodies on and off the trucks. The Fitch scheme worked well; his Motor Terminals Company appeared to be a promising business venture. "Each truck made twenty-six trips in a ten-hour day," noted transportation historian John H. White Jr. "All together they were moving 660 tons a day and reckoned they could transfer almost double that amount if required to do so. The system reduced the platform space, and damage and loss claims all but disappeared." Although various refinements and modifications followed, including special containers for raw milk, steam railroads showed only modest interest. Yet, the Fitch firm continued to function until the 1960s. Observed White, "It seems ironic that just as the container was becoming a major force in American transportation, Cincinnati's pioneer operation expired" (White 2003, 43).

Long after Motor Terminals Company moved its first revenue load, Malcolm McLean, a motor carrier executive from North Carolina, promoted the idea of simply separating the road wheels from the truck trailer. This approach, which he began to promote in the 1950s, allowed movement of freight by highway, railroad, and ship. After considerable obstacles, but largely nontechnical ones in nature, McLean successfully converted an old oil tanker into what he called a "seagoing tractor." In the spring of 1956 his *Ideal X* with a load of fifty-eight containers sailed between the ports of New York and Houston, Texas. Two years later the Matson Navigation Company launched containerization between San Francisco, California, and Honolulu, Hawaii. While the concept looked encouraging, railroad involvement remained limited, largely due to cost considerations. If intermodal operations received funding and technical assistance, it was earmarked for expansion of TOFC. But that would ultimately change.

Companies that spearheaded this new wave of intermodal operations would be the Chicago & North Western Railway (C&NW) and Union Pacific (UP). (Indeed, this close service relationship led UP to acquire

C&NW in 1995.) In the late 1970s, the progressive management at C&NW sensed the bright promise of container traffic and began dispatching COFC trains, with their standard cargo boxes, between Chicago and the West Coast in conjunction with UP through an interchange at Fremont, Nebraska. "This [containerization] is really a railroad revolution," explained a C&NW official to the author. "It is hard to believe that one time goods were packed in crates, bags and barrels that were loaded, unloaded, loaded and even unloaded again." This container service worked well and became an important revenue source for both companies.

Then in 1984, the latest twist in COFC occurred: *double-stacks*. These trains began to move between the Midwest and the Far West via C&NW-UP-Southern Pacific, and soon these carriers added a Consolidated Rail Corporation (Conrail) connection at Chicago for the East Coast, forging the first transcontinental double-stack container operation. The American President Lines, a leading freight forwarder, played a key role by investing heavily in double-stack cars built by the Thrall Car Manufacturing Company of Chicago Heights, Illinois. Loaded in Pacific Rim nations, containers arrived at West Coast ports where giant cranes transloaded them from specially designed ships onto low-to-the-rail, lightweight flatcars. This relatively inexpensive

Cargo containers have become an important part of contemporary freight movements. In July 2004 an eastbound Union Pacific train near Echo, Utah, has in its consist several "double-stacks." The contents of these containers likely originated in a Pacific Rim nation. © William W. Kratville.

Although the trailer-on-flatcar (TOFC) and container-on-flatcar (COFC) have become commonplace, just a few carriers, most notably Norfolk Southern, have opted for the technology of the RoadRailer (above). Still, the equipment works well; these combination truck trailers/train cars have demonstrated their reliability and practicality. Detail of the RoadRailer is shown on the next page. Courtesy of the Norfolk Southern Corporation.

rolling stock permitted a train to carry twice as many containers with only a 30 percent increase in weight compared to conventional COFC trains. A simple pin-linkage system secured the top container to the bottom unit. At their destinations fixed cranes, monster straddle machines, or some other advanced lifting devices placed containers on an ocean-going vessel for European ports (the United States served as a "land bridge") or on a truck chassis for domestic delivery. Less tonnage, however, made the reverse trip, although thousands of empty containers did. Everything functioned smoothly. "Double-stack technology has enabled the C&NW to move 80 percent more loads per train

at the same cost as traditional piggyback service," happily reported management in its *1992 Annual Report.* "The average revenue per car earned for double-stack movements is lower than our other business groups, but profit margins are higher because of the inherent efficiencies of the operation."

Technologically speaking, approaches to intermodal service did not move neatly from piggyback to containerization. Rather, other alternatives emerged, including "Flexi-Van" and "RoadRailer." In both cases, these intermodal schemes were tied directly to a single carrier: New York Central (NYC) with Flexi-Van, and Norfolk Southern (NS) with RoadRailer.

In the mid-1950s executives of the NYC wished to convert their high-speed passenger railroad into a fast-freight operation. These men correctly foresaw the decline of long-distance passenger travel, being acutely aware of continual improvements in commercial aviation technology. Since the company had some clearance bottlenecks on its main lines into both Boston and New York City, it sought an innovative way to solve these problems. The result was Flexi-Van, designed to use low-profile equipment and to originate and terminate intermodal shipments at more remote and hence inexpensive terminal sites.

The prototype Flexi-Van car worked as intended. "The . . . [truck] van backed at right angles to the flatcar until its bottom rails lined up with the car's turntable mechanism," explained a historian of intermodal development. "The driver then unlocked a pin in the demountable bogies, and backed the van off the wheels and into slots on the turntable. When the van was fully engaged on the turntable, a pin locked it in place. Using a push rod mounted on the Commando hostler unit, the driver engaged the van, pushed it 90 degrees until it paralleled the flatcar and locked into place on the car" (DeBoer 1992, 63). This engineering was basic but effective.

Although NYC officials believed that Flexi-Van performed well, this particular intermodal technology was not widely copied. A few railroads flirted with the system, but it would remain almost exclusively an NYC operation. Then the company improved its main line clearances and embraced conventional TOFC. That emphasis, and then in 1968 NYC's merger out of economic necessity with the Pennsylvania Railroad, doomed Flexi-Van. It became but another example of a rejected technology.

Although Flexi-Van fizzled, RoadRailer turned out much differently. In the late 1970s a private concern, the Bi-Modal Corporation, created a prototype of what it smartly called RoadRailer, namely, fixing a truck trailer with a set of single-axle rail wheels that were located behind the conventional tandem highway axles. Thus, a truck trailer could travel by rail without a flatcar.

In reality, RoadRailer was a modernized version of the earlier, albeit disappointing *Railvan*. In the late 1950s the Chesapeake & Ohio Railroad (C&O) introduced a creative way to maximize "payload" by substantially reducing tare weight. Technical bugs and the tendency of these lightweight Railvan units to derail and then rerail themselves prompted the C&O to exit the experiment for its mail and express service between Detroit and Grand Rapids, Michigan. Notwithstanding the Railvan experience, the RoadRailer concept was pushed ahead. Although some encouraging signs developed with operations on several railroads, including Conrail, it was in 1986 when RoadRailer service started in a big way on Norfolk Southern.

By the beginning of the twenty-first century Triple Crown Services, Inc., an NS subsidiary based in Fort Wayne, Indiana, and located at the hub of three NS main lines, supported RoadRailer operations on both NS and Canadian National rail lines to such major market centers as Atlanta, Chicago, Dallas, Jacksonville, Kansas City, Montreal, and Toronto. The technical systems consist of three principal components. First there is a specially designed, high-strength, "unibody," lightweight 53-foot-long trailer, featuring slack-free couplers that securely connect trailers when in train formation without transmitting potentially damaging, rocking, or bouncing action from one trailer to another. Then there are the high-speed rail bogies that carry trailers on the rails and provide a suspension system with shock-absorbing steel and rubber springs. And finally the RoadRailer offers what management calls the "CouplerMate," namely, the connection between conventional rail equipment and RoadRailer units. Outside of North America where train lengths are usually much shorter, groups of RoadRailer trailers can be placed in any position. The company, too, boasts that the 12-inch coupling distance between trailers during rail travel prevents the opening of cargo doors, thus eliminating theft or vandalism. With essentially no tare weight, these RoadRailer trailers can go anywhere and do anything that conventional truck trailers can do and, of course, they can ride directly on the rail. The era of *hybrid* rail-highway vehicles has come of age. For railroads it provides an effective way to win business back from truckers and to satisfy the "just-in-time" production schedules of automotive and other manufacturers. No one disputes that intermodal trains play a vital role in the market-supply chain. Therefore the three main technologies of intermodal service—TOFC, COFC, and RoadRailer—have contributed greatly to the phenomenon of railroad Rebirth.

A BETTER RAIL*ROAD*

As intermodal trains race between terminals, they can do so not only because of more powerful diesel-electric locomotives and lightweight, modern rolling stock, but also because of a better track structure. This remains a vital aspect of railroading. Obviously, even the best automobile design in the world cannot improve upon a bumpy ride if the road that it is driven on is poorly maintained. And on railroads no car, wheel, truck, or cushioning design will ever eliminate the need for a well-run maintenance-of-way program to keep the track and all its structures in top-notch condition. In 1912 this practical philosophy was aptly stated when Howard Elliott, president of the Northern Pacific Railroad, told delegates of the Tri-State Grain and Stock Growers' Association that "you cannot have a good railroad without good track and good equipment, and good men to maintain and operate that track and equipment" (*The Truth about the Railroads* 1913).

Since Elliott spoke, much has changed. The historic physical symbols for maintenance-of-way employees (trackmen) might be the spike mall and the rail-alignment bar or "idiot stick." Then, during the latter half of the twentieth century, mechanization of track maintenance, at times later assisted by computerized data, revolutionized this dimension of railroading. Surely one factor in this trend involved the shift to the 40-hour week for maintenance-of-way employees in 1949 that promoted railroads to seek greater mechanization as an economic necessity. And so companies purchased power adzes, wrenches, spike drivers, and more.

By the 1970s trackmen had at their disposal an array of mechanical equipment. Gangs worked with highly automated tie machines (which quickly removed old ties and installed replacement ties), ballast-cleaning and ballast-tamping equipment, and rail-alignment devices. By the 2000s the latest bells and whistles provide more computer control to reduce the possibility of human error. Heavy equipment, ranging from large cranes that travel from highway to railway to monster earth-moving machinery, also makes line construction and right-of-way rebuilding faster and less labor intensive.

Maintenance-of-way practices changed in other ways. No longer do these men of the steel rail eat and sleep in old, hot, cold, or drafty wooden bunk or "outfit" cars and reach the worksite on foot, on small hand-pump cars, or on later motorized "speeders." Rather they stay in climate-controlled mobile trailers (some mounted on railroad flatcars) or simply overnight in nearby motels, dine in local restaurants, and travel by car, SUV, or four-wheel-drive truck (often equipped with hydraulic flanged wheels) to the job site. In recent years the latter set of practices, in fact, has become prevalent since railroads have turned more to outside contractors to manage their physical plants.

Just learning about what needs to be done has undergone enormous change. Although for decades railroads have used track-detector equipment to check on the overall condition of rails, including flaws, wear, and track geometry, high-tech information and communication methods have wrought additional improvements. At the beginning of the twenty-first century, inspectors can take automated inspection data, download it into a Palm

Pilot–based system, and use a global positioning satellite receiver to follow up behind automated inspection cars. Personnel are able to use these technologies in their primary reports. Instead of writing out paper forms and then transferring the information onto an office computer, this data can enter the network while employees are out in the field, resulting in greater speed, efficiency, and reliability.

Although the lowly crosstie remains a basic railroad component, it, too, has been altered. Although railroads annually acquire millions of chemically pressure-treated hardwood ties, they also install tens of thousands of ties made of reinforced concrete, especially in wet and humid areas where rot and insects cause repeated damage. It is understandable that the Jacksonville-to-Miami Florida East Coast Railway became an industry leader in the concrete alternative.

Concrete ties are more than rot and insect resistant; now they can be scientifically contoured. This has resulted in the placement of support materials where they are needed most, namely, under the rails, and relatively little in the middle where weight requirements are limited. Conventional wood ties, however, do not lend themselves to such creative shaping. Moreover, railroad maintenance-of-way experts have found that track with concrete ties holds alignment well. And using clean, high-quality ballast when concrete ties are first installed helps to assure good surface performance.

Still, wood ties continue to dominate the North American market, holding more than 90 percent of market share. That explains why wood-product engineers and others seek to improve overall quality. In recent years, a "wood composite" tie has appeared. This alternative to standard hardwood products offers some attractive features, including ease and flexibility of installation and the use of smaller, younger trees for production.

Perhaps the tie for the twenty-first century may be neither manufactured from concrete nor produced from wood ingredients. Since the mid-1990s research has been ongoing on fashioning plastic composite ties, and they appear to offer several potential advantages. Plastic composite ties are resistant to decay and insect damage without additional treatment, and installation is similar to that of wood ties. Moreover, these plastic products also provide increased lateral track stability when the sides and bottoms are properly roughened, and they are recyclable. Union Pacific, that perennial industry leader in the application of the latest technologies, has experimented successfully with these plastic composite ties. Steel ties, on the other hand, used in a few places domestically, have not found widespread usage. Still, some railroads, mostly abroad, find that these all-metal ties provide good gauge restraint under heavy loads and, of course, have great durability.

COMMUNICATIONS AND REMOTE CONTROL

Maintenance-of-way personnel have kept in touch with one another with ever-improved pieces of technology. Earlier it might have been land-line telegraph and then telephone connections, later two-way radios, and in recent years cellular phones. Railroad workers, of course, require reliable and abundant communication facilities. Although the historic network of telegraph and telephone equipment (including so-called carrier telephone lines where more than one message could be handled) worked reasonably well, these methods of information transfer have had modest capacity. They also were labor intensive, requiring full-time maintenance crews and additional support staff. And storms, whether ice, sleet, snow, or wind, could cause major destruction, even forcing suspension of rail operations.

In the early 1960s an important betterment in the field of railroad communications emerged: some railroads embarked on installing microwave radio transmission systems. Generally speaking, this replacement technology involved radio communications that used frequencies above 1 gigahertz, and consisted of radio transmitters and receivers capable of being modulated simultaneously by various voice channels (twelve to 600 in railroad use). Once carriers had installed their primary microwave networks, they expanded operations by adding lower density systems that allowed them to achieve nearly universal coverage of their service territories.

These upgraded communication networks held numerous advantages. For the most part, they possessed greater capacity, featured higher voice quality, and required less maintenance than wired facilities. Furthermore, microwave communications were usually not susceptible to adverse weather conditions. The rate of service reliability was nearly 100 percent, a shining performance record.

The Southern Pacific Railroad (SP) did much to spearhead this communication revolution. By the early 1970s the firm launched a subsidiary, Southern Pacific Communications Company (SPCC). Not only did this new affiliate handle railroad business with its state-of-the art communications networks, but it also won approval from the Federal Communications Commission (FCC) to acquire commercial customers. Thus SPCC entered the competitive world against American Telephone & Telegraph (AT&T) and Western Union for interstate telephone and message services. But in time the financially strapped SP sold off SPCC, and ultimately it became the much-touted Sprint unit of MCI.

From the Gestation years onward, railroads have been involved in some form of train control and dispatching. The objective has remained the same, namely, to get the job done safely and with the greatest efficiency in terms of moving trains over the road with the reliability and speed that customers demand. For decades a chief dispatcher and assistants used telegraphy to get in touch with train crews. Orders went to open telegraph stations where operators copied messages, including "meets," and gave them to head-end and rear-end crews by writing the brief order on onionskin paper called "flimsies" and then attaching them to hoop-type sticks for the fireman and brakemen to scoop up "on the run." Orders also were given to crew members when they left the terminal or made regular stops. Similarly, station agents adjusted train-order boards, and towermen threw signals and switches based on directions from the dispatchers' office. Subsequently, telephones, automatic block signals, and centralized traffic control made for better train management. But in recent years computers, microwave radios, and related technologies have dramatically altered dispatching. No longer does a dispatcher, wearing the traditional green eyeshade, systematically mark with his fountain pen the blanket-size train sheets to remain aware of the location of all movements, whether a local freight performing switching chores or a through, fast "red ball" main line merchandise train. And no longer does he need to remember the places where maintenance-of-way employees are at work or where there are "slow orders" because of "soft," dangerous track, derailments, or some other hazard.

In 1988 a historic event in the development of railroad train control work took place in Jacksonville, Florida. CSX Transportation Company, the sprawling giant carrier made possible by the merger somewhat earlier of the Chessie System with the Seaboard Coast Line, opened what it called the Dufford Control Center. No large railroad had ever before brought together *all* train control functions in a single location. And this facility was state of the art. By the use of computers with specially designed software, developed and installed by Union Switch & Signal Company, and the latest radio equipment, dispatchers were able to keep track of movements on this vast interregional system. Direct connection with the U.S. Weather Bureau, for example, allowed Dufford Control Center personnel to learn immediately of lightning strikes, heavy downpours and flash flooding, and ice and snow storms.

Other major railroads, too, rapidly incorporated these types of technologies for their train-control and -dispatching operations. Most followed the CSX model for centralization. A variety of specialized adjustments became common. The Santa Fe, for one, not only paid attention to weather reports, but its Systems Operations Center also established a direct link to the University of Southern California's seismology office, providing immediate notification of any tremors in earthquake-prone California. This allowed the railroad to make inspections quickly in case of a quake registering 5.0 or higher on the Richter scale.

As might be expected, a refinement process in this overall dimension of modern railroading has been ongoing. Contemporary railroads are moving toward total "asset utilization." The day is rapidly approaching when all the information a dispatcher needs to manage his territory and make it fit seamlessly into a fluid, schedule operation will be a reality. What is happening is that the dispatcher, the first point of contact with the physical railroad, is becoming a transportation planner. With the development of more sophisticated equipment, so-called movement planning will be made possible by the application of "smart software" (complicated algorithms) that will automate the meet-pass process of trains that presently occupies much of a dispatcher's attention in what remains a mostly manual operation. The dispatcher then can fine-tune operations, adapting to different traffic situations quickly and safely.

Although casual observers might not sense the communications revolution, they are surely aware that the freight train is without a traditional feature—the caboose. Like the steam locomotive, the "little red caboose behind the train" has nearly vanished. In the 1990s most railroads retired their fleets of cabooses, contending that these pieces of rolling stock were expensive and unnecessary. Today an electronic black box or "end-of-train device" (ETD), mounted on the rear coupler of the last car and hooked into the train-brake line, provides a suitable substitute. A flashing red strobe light in the ETD shows to operating personnel and trackside watchers the final car, and a radio sends the air pressure in the brake lines and motion information to the head-end personnel, informing them if there are problems. With their ETD readouts, engineers can handle mile-long trains more smoothly and safely, with less slack action. Companies, both in North America and throughout the world, discovered that these ETD devices are rugged and that they deliver continuous operation in extreme environments, ensuring few failures and low maintenance. Moreover, railroads found that there were other reasons why there was no longer a need for the brakemen/conductor to ride in the caboose; the largely simultaneous development of dragging equipment and hotbox detectors can keep the head-end crew fully informed on these potentially dangerous situations.

For decades, railroad personnel have thought about a revolutionary approach to routine switching chores in yards and terminals. In the late 1950s, for example, Gregory Maxwell, president of the Terminal Railroad Association of St. Louis (TRRA), toyed with the use of a remote-control device to allow a brakeman to maneuver a locomotive while breaking or making up a train. But because of union opposition, Maxwell abandoned what

"I thought was an excellent way to improve operating efficiencies, to bolster worker safety, and to cut costs and hence boost profits on the TRRA. It was obvious to me that men in the cab or on the ground are idle much of the time, and that's not good." During his conversation with the author, Maxwell added, "I learned again sadly that you can't roil organized labor."

A few years before Maxwell experienced his interest, remote-controlled locomotives (RCL), albeit small in size and horsepower, began to appear on *private* industrial railroads. It did not take long before they came into common use at breweries, mines, refineries, steel mills, and similar facilities. Such automated activities did not cause union complaints, since nonrailroad labor employees worked at these sites. And, significantly, these operations were successful in all ways.

In the 1960s a variation of the developing technology for remote-control trains began to appear. This was employment of radio-remote locomotives that eliminated manned or "helper" engines. These devices did away with the helper crew and enabled extremely heavy trains to run across long hill-and-dale districts. Locomotive engineers liked this technology because it gave them sole control of the train rather than requiring them to wonder what the helper engineer would do. While early radio-remote technology was fraught with problems, ultimately the technology became reliable.

In the 1990s the big breakthrough for remote-controlled locomotives took place. Railroads in Canada, Germany, New Zealand, and several U.S. "regionals" and shortlines began to embrace a more fully developed "automatic" technology. By the end of the 1990s the Canadian National Railway, for one, had placed about three-quarters of its flat-yard switching operations and all of its hump-yard engines under remote control. And this technology has allowed roads, mostly shortlines, to go after low-density customers, further enhancing the equipment's economic value. In 2003, the head of the 155-mile Indiana Rail Road (IRR), a dynamic Midwestern shortline, noted that because of RCLs, his company can profitably handle a few carloads that move only short distances. "IRR doesn't use RCLs everywhere, only where it makes sense to do so," a practice common to this type of customer-sensitive carrier (Kube 2003, 46).

Remote-control technology is largely the same worldwide. A major producer, CANAC, Inc., based in Montreal, Quebec, produces what it calls the "Beltback," equipment that allows a yard operator to control driverless, microprocessor-equipped switching locomotives by using what is known in the industry as an operator control unit (OCU). There are seven speed and five brake settings, and it works like this: the operator selects a speed, and the Beltpack supplies just enough power or brake to maintain that speed. It automatically accounts for such variables as track grade and train mass. Furthermore, a so-called pitch-and-catch system permits control to be transferred between operators at opposite ends of the train. Thus one operator can "pitch" or transfer control of the locomotive's movement to a colleague who "catches." This allows each operator to observe switching activities from different vantages. Another feature of the OCU is a predetermined coupler speed. The onboard computer is programmed to override certain unsafe commands and thereby preventing equipment damage. Moreover, the Beltpack system is relatively easy to learn and to maintain.

Ultimately, the goal of railroad management is to operate all or nearly all unmanned trains. Again, cost considerations are paramount; fewer jobs mean greater profits. For some these might be future fantasies, but already there have been examples. Between 1968 and 2002, the 15-mile Muskingum Electric Railway (MER), located in southeastern Ohio, moved hopper cars of coal with fully automated trains. The MER functioned uneventfully until the mining operations closed. Recently the Quebec, North Shore & Labrador Railway began operating seven unmanned twenty-car shuttle trains between iron-ore loading points and a processing plant from 5 to 7 miles apart. Automation equipment, supplied by General Electric Transportation Systems, made this service possible, but even GE, like others, is uncertain if the larger railroad industry is ripe for what would be truly radical change.

Even if a railroad company has not embraced any form of RCL technology, it likely has employed (or will employ) a major emerging revolution in the braking of freight trains. In recent years electropneumatic brakes (ECP) have become increasingly popular worldwide, especially on long, heavy-unit trains. Rather than traditional AB "mechanical" devices, ECP units offer an improved way to control cars. By attaching to each car an axle generator power source or a current pneumatic power unit, namely, a turbine driven by brake pipe pressure, these ECPs provide the engineer with faster stopping capabilities. By using radio technology, a signal can be sent from the locomotive almost instantaneously to the brake system and a computer-recorded response follows as rapidly, indicating to the engineer if everything is functioning properly.

HIGH-SPEED TRAINS

If there are poster children for the railroads of tomorrow, they surely appeared in the closing decades of the twentieth century with the debut of

It did not take long for Amtrak, which in the early 1970s acquired ownership of the high-speed electrified corridor of the Penn Central Railroad between Hartford, Connecticut, and Washington, D.C., to upgrade the rolling stock, including electric locomotives. An Amtrak electric train, pulled by two AEM-7-type electric locomotives, roars through Connecticut. Courtesy of the author.

long-distance Bullet trains in Japan, FasTrains in Sweden, TGVs in France, and ICEs in Germany. These very high-speed conventional trains have caught the public's fancy, and for good reason; they offer practical and attractive alternatives to air and other land transport.

In the mid-1950s the direct origins of these popular fast trains arguably began in France. The French National Railways (SNCF) used a recently electrified stretch of main line track to dispatch trains that sped at more than 150 miles per hour. This impressive demonstration piqued greater interest, and additional testing followed. But SNCF went beyond its original rolling stock; these new, experimental trains featured regeared locomotives, specially designed pantographs, a rounded trail section (to reduce drag), and more electric power.

By the early 1960s SNCF officials had decided on an important new concept for the passenger railroad: they sought to combine high speeds with *steep grades*. Their electrically powered trains would speed over a gentle roller-coaster terrain. Rather than the 1 percent or perhaps 2 percent dominant right-of-way grades, considered steep in conventional operations, the line would have grades that reached about 4 percent. Such a construction practice would allow more flexible and cheaper routing of new lines, and this trackage would be dedicated solely to high-speed passenger trains. For safety purposes, there would be no public or private road crossings and only like trains would be permitted to use the line.

French railroad personnel were not certain about motive power. The debate centered on whether to use electric or gas-turbine locomotives. Both were tried, but by the mid-1970s the SNCF accepted the electric option (three-phase AC induction motors). The reason involved what was believed to be the less economical gas turbine, in part because of the oil crisis of the time. Furthermore, all-electric traction offered better integration of all components of the railway, namely power supply, signaling, and control. Experts also solved the problem of picking up power from overhead contact wires at extremely high speeds.

The French creations truly look like trains of the future. The most striking aspect was their aerodynamic styling of the nose section of the lead unit. Yet there was more that the casual observer might not recognize. The train set was articulated; the cars were not merely coupled together, but rather were semipermanently attached to each other, with the ends of two adjacent cars resting on a common two-axle truck. This took place because designers sought to create extremely lightweight trains, and reducing the number of axles helped to achieve that engineering objective. But articulation offered more: it reduced interior noise levels; allowed clean, quiet passage from one car to another; and prevented the kind of jackknifing that usually happens with conventional trains in a collision.

This cutting-edge equipment featured a much-talked-about innovation. The system employed *in-cab* signaling. The initial and subsequent high-speed rail routes do not use line-side signals, being too difficult for the operator to read at speed. All signaling information is transmitted to the train through the rails, and appears on easy-to-see indicators in the cab. In general, the train sets rely heavily on computerized data.

In 1981 the high-speed French rail project grabbed worldwide attention. In February a sleek orange passenger train, dubbed the TGV or le Train à Grande Vitesse (high-speed train), reached a speed of 380 kph (236 mph) while on a test run near Lyon. Then seven months later TGV trains, French designed and French built, flashed across the countryside between Paris and Lyon on their initial revenue runs. Two years later, when the entirety of the new line was finished, passenger travel time between the two cities dropped to only two hours. Thus began the long tradition of high-speed ground transportation in France, and by the 1990s the company had three TGV lines in operation—TGV-Sud-Est, TGV-Atlantique, and TGV-Nord—and recently it placed into service double-decker cars for greater passenger-carrying capacity. Moreover, TGV technology became a contender in various export ventures, including China, South Korea, Spain, and Taiwan. And TGV trains presently serve many parts of Europe—namely, Belgium, Great Britain, Germany, Italy, the Netherlands, and

As the twenty-first century begins, high-speed intercity trains race through the European countryside. These electric-power passenger units take advantage of the latest streamlining designs and electric-power systems. The top German Dutch Intercity Express (ICE) is captured near Breukelen, Holland, on the Amsterdam-Frankfurt AM main line. The bottom French-Belgian-Dutch *Thalys* train is seen near Dordrecht, Holland, on its Amsterdam-Paris run. Courtesy of John Krijgsman.

Switzerland—and routinely roll down ribbons of rail at almost 190 mph. A new "golden age" of passenger rail service seems at hand.

THE FUTURE

While the future may well see the wider use of high-speed trains, even trains operated on the maglev (magnetic levitation) principle, and repeated improvements in a variety of technologies, especially computer-related ones, it appears probable that these improvements will not be the only means to enhance productivity and service. It is reasonable to believe that adding additional horsepower into locomotives is not too likely; after all, 6,000 horsepower is more than enough for the longest trains. And, too, it is unlikely that freight cars will become much bigger or heavier. Admittedly, recent trends have been to make equipment lighter by using aluminum and special metal bodies but adding freight-carrying capacity, yet not changing the overall gross weight on the rails. Indeed, contemporary railroad industry experts generally do not see technology alone as the *sole* answer for better railroads. The feeling exists that better equipment utilization, planning, and use of infrastructure are all highly important for future railroad enhancements.

Still, it cannot be overlooked that in the not-so-distant future, high-speed passenger trains will surely become part of the North American transportation network. Why will that happen? The answer to that "why" has much to do with the deterioration of the Interstate Highway System, the expense and difficulty of expanding these ribbons of concrete, the clogging of the airways with delays and other hassles at airports, and ongoing concerns about energy consumption and vehicular pollution. High-speed/maglev could act as a feeder to major hub airports, eliminating numerous short flights. For distances of up to 500 miles, a high-speed ground transportation system could function much better than airplanes. Moreover, Americans have always had an obsession with speed, speed on the ground as well as speed in the air. Rates of greater than 125 miles per hour, presently reached on Amtrak's Northeast Corridor, will not be enough. Unmistakably, technological advances, whether in France, Germany, or China, point to an exciting, fast future of guided ground transportation.

Some "futurists" believe that maglev will be *the* train of the twenty-first century. The concept is exciting. The big difference between a maglev train and a conventional train is that the former lacks an engine. Rather, the basic principal is that magnets have opposite poles that attract and like poles that repel and that the magnetic pull is *temporary*. It is the magnetic field created by electrified coils in the guideway walls and the track that combine to propel the train. Specifically, a magnetized coil, which runs along the elevated guideway supported by pylons, repels large magnets on the undercarriage of the train, allowing the vehicle to levitate between 1 to 10 centimeters (0.39 to 3.93 inches) above the guideway. Once the train is in this levitated state, power is supplied to the coils located within the guideways walls to create a system of magnetic fields that push and pull the train. The electric current, which is supplied to the coils, is constantly alternating to reverse the polarity of the magnetized coils, and this change in polarity in the magnetic field behind the train provides additional forward thrust. Since maglev trains float on a cushion of air, the lack of friction and the aerodynamic design of the trains allow incredible speed, more than 500 kph (310 mph), considerably faster than the latest TGV units. Moreover, energy consumption is low, about half that required for an electrified railway, and the system is environmentally friendly. For one thing, the space required for the guideway is less than half that of a highway, and somewhat less than that of a conventional electrified railway. And since maglev trains are capable of negotiating maximum grades of more than 10 percent, the need to displace large amounts of land for tunnel construction is significantly reduced.

As with any radical type of transportation technology, there are opponents and detractors. A common objection is that the maglev can merely go from point A to point B, and would be unable to link physically with conventional railroad trackage. Moreover, while there are environmentally attractive features, there are complaints about the ugliness of the structures needed to create such transport arteries; high amounts of unattractive protruding concrete works would be mandatory and would surely clash with the natural surroundings.

Proponents of maglev, nevertheless, have great expectations. They strongly believe that ultimately these trains will become a suitable alternative to airplanes, especially for distances of less than 1,000 miles. Prototypes have been developed in China, Germany, and Japan, and it is conceivable that there will be a magnetic wave of the future, although perhaps modest, being limited to specific high-population corridors and between large "hub" airports and center cities.

What about the new century? People and goods must always be moved, and surely conventional railways are not likely to disappear in the near term or perhaps ever. Steel-on-steel offers much; no futurist anticipates the demise of this means of transport. Yet, understandably, the quest for a practical technology continues, with its triumphs and disappointments, just as it has since the dawn of the Railway Age.

Appendix: Steam Locomotive Types

Steam locomotives are usually referred to by their classification in the Whyte System of Engine Classification. At the turn of the twentieth century an Englishman, Frederick Methvane Whyte, used a series of numbers to show the number of pilot-truck or leading wheels, drivers, and trailing-truck wheels on each type of steam locomotive. For example, the early "American" type is a 4-4-0, meaning that it has a four-wheel pilot truck, two pairs of coupled drivers, and no trailing wheels. Each group of drivers on articulated and multicylinder nonarticulated locomotives is shown separately. Therefore, the Union Pacific's *Big Boy* is a 4-8-8-4.

Type	Name
0-4-0	*Four-wheel switcher*
0-6-0	*Six-wheel switcher*
0-8-0	*Eight-wheel switcher*
0-10-0	*Ten-wheel switcher*
0-10-2	*Union*
2-4-2	*Columbia*
2-6-0	*Mogul*
2-6-2	*Prairie*

2-8-0	*Consolidation*
2-8-2	*Mikado or MacArthur*
2-8-4	*Berkshire*
2-10-0	*Decapod*
2-10-2	*Santa Fe*
2-10-4	*Texas*
4-4-0	*American*
4-4-2	*Atlantic*
4-4-4	*Jubilee*
4-6-0	*Ten-wheeler*
4-6-2	*Pacific*
4-6-4	*Hudson*
4-8-0	*Twelve-wheeler*
4-8-2	*Mountain*
4-8-4	*Northern*
4-10-0	*Mastodon*
4-10-2	*Southern Pacific*
4-12-2	*Union Pacific*
0-6-6-0	*Mallet compound*
2-6-6-0	*Mallet compound*
2-6-6-2	*Mallet compound*
2-6-6-4	*Simple articulated*
2-6-6-6	*Allegheny*
0-8-8-0	*Mallet compound*
2-8-8-0	*Mallet compound*
2-8-8-2	*Mallet, simple articulated*
2-8-8-4	*Yellowstone*
2-8-8-8-2	*Triplex*
2-8-8-8-4	*Triplex*
2-10-10-2	*Mallet compound*
4-4-4-4	*Four-cylinder nonarticulated*
4-4-6-4	*Four-cylinder nonarticulated*
4-6-6-4	*Challenger*
4-8-8-4	*Big Boy*

Glossary

AB Brake. The standard freight-car brake system that consists of the AB control valve, brake cylinder, auxiliary and emergency reservoirs, train line, and associated parts. The system allows rapid series brake applications on each car in a train and is controlled from the locomotive cab, replacing after World War II the Type K brake.

Air Brake. The general term used to describe the braking system used on most railways that operate in North America.

Air Conditioning. Generally the simultaneous adjustment of atmospheric conditions, chemical as well as physical, within an enclosed space, to suit the requirements of the purpose for which that space is used, regardless of variations in natural atmospheric conditions.

Alternating Current (AC). An electric current that reverses its direction at regular intervals.

Articulated Car. A car created by uniting two or more railcars to form a single unit.

Automatic Train Control (ATC). A trackside system working in conjunction with equipment installed on the locomotive, so arranged that its operation will automatically result in the application of the air brakes to stop or control a train's speed at designated restrictions should the engineer not respond. ATC usually works in conjunction with cab signals.

"B" Unit. A diesel unit without a cab and without complete operating controls.

Belpaire Boiler. A popular steam locomotive boiler type that features longitudinal bulges on each side near the top that results in a flat upper surface, a shape that creates enhanced structural strength and offers a greater surface area for heat transfer.

Bogie. The running gear of a highway semitrailer that may be removable or longitudinally adjustable. Also the European railway term applied to railway freight and passenger car trucks.

Brake Cylinder. A steel cylinder attached to the body frame or truck frame of a car or locomotive containing a piston that is forced outward by the compressed air to apply the brakes. When air pressure is released, the piston returns to its normal position.

Brake Shoe. A block of friction material formed to fit the curved surface of the tread of a wheel, and riveted or otherwise bonded to a steel backing plate having provision for quick and positive securement to the brake head. Brake shoes can be made of cast iron or of a high-friction composition material.

Camelback Locomotive. Differing from a regular steam locomotive, a camelback engine positioned the engineer near the front and the fireman at the back close to the coal supply.

Canadian Gauge. Used in both Canada and the United States, the width of the track gauge measures 5 feet, 6 inches.

Catenary. On electric railroads, the term identifies the overhead conductor that is contacted by the pantograph or trolley and its support structure.

Centralized Traffic Control (CTC). A traffic control system that allows a dispatcher to control train movements by automatically setting signals and switches.

COFC. An acronym for "container-on-flatcar." A type of rail freight service that involves the movement of closed containers on special flatcars.

Combination Depot. This small-town or "country" depot generally consists of three sections: center office, end waiting room, and baggage-freight storage section.

Compound Locomotive. A steam locomotive that can reuse steam by having both high-pressure and low-pressure engines. Ideally, a compound locomotive realizes more power from its steam than does a simple or high-pressure engine.

Compound Rail. An early rail type that consisted of two or more iron pieces that were either bolted together or interlocked like pieces of a jigsaw puzzle.

Compromise Car. A common freight car of the nineteenth century that featured opened-barred side doors and end ventilator doors and could be used for hauling both general merchandise and livestock.

Container. An independent structural unit, either open or fully enclosed, designed for the intermodal transport of commodities.

Continuous Welded Rail (CWR). A fabrication process whereby sections of steel rail are welded together, providing for smoother riding track and reducing maintenance costs.

Demonstration Period. The first several decades of the Railway Age (1830s–1850s) before rail, rights-of-way, stations, motive power, and rolling stock became heavily standardized.

Depot. *See* Station.

Diesel Engine. An *internal* combustion engine invented by Rudolf Diesel that differs from other internal combustion engines because its compression is high enough to cause combustion without the need of a spark for ignition.

Diesel-Electric Locomotive. A locomotive in which power developed by one or more diesel engines is converted to electrical energy and delivered to the traction motors for propulsion.

Diesel-Hydraulic Locomotive. A locomotive in which power developed by one or more diesel engines is delivered through a hydraulic transmission to the driving axles by means of shafts and gears.

Direct Current (DC). An electric current that flows in one direction only.

Drivers. The powered wheels on a steam locomotive.

Dynamic Braking. A term used to describe a method of train braking whereby the kinetic energy of a moving train is used to generate electric current at the locomotive traction motors, which is then dissipated through banks of resistor grids in the locomotive car body.

Dynamometer Car. A car equipped with apparatus for measuring and recording drawbar pull, horsepower, brake pipe pressure, and other data connected with locomotive performance and train-haul conditions.

Electric Locomotive. As distinguished from a gas-electric or diesel-electric locomotive, a self-propelled vehicle, running on rails and having one or more electric motors that drive the wheels. The motors obtain electrical energy either from a rail laid near to, but insulated from, the track rails or from a wire suspended above the track. Contact with the wire is made by a pantograph or trolley wheel on the end of a pole mounted on top of the locomotive.

Electro-Pneumatic Brake. A braking system used on high-speed electric passenger trains. Brakes are applied and released on each car through the action of electro-pneumatic valves energized by current taken from contacts on the motorman's brake valve and continuous train wires. Brakes can be applied instantaneously and simultaneously with this device, eliminating undesirable slack action and providing better control of train speed.

End-of-Train Device (ETD). An electronic black box placed on the last car and connected to the train brake line, providing crew members with data on air pressure and other dimensions of train movement.

Erie Gauge. First introduced by the New-York & Erie Railway and subsequently used by several other, mostly eastern carriers, the width between the rails measures 6 feet.

Flange. *See* Wheel Flange.

Flues. Sometimes known as "fire tubes," these metal pipes are located in the boiler of a steam locomotive and allow hot gases to pass on their way to the stack, and in the process convert the surrounding water into steam. At times, flues become clogged and must be bored out or replaced.

Gas-Turbine Electric Locomotive. A locomotive in which power is developed by a gas turbine that drives electric generators supplying current to electric traction motors on the axles.

Geep (GP). Introduced in 1949, this popular "general purpose" diesel-electric locomotive, manufactured by the Electro-Motive Division of General Motors, handles both freight and passenger train assignments and became a model for later locomotive development.

Gravity Railroad. An early means to propel railway cars using the force of gravity to move minerals, lumber, or the like, usually from the point of origin to a body of water. Pulleys and wheel brakes controlled speed.

Hand Brake. A device mounted on cars and locomotives that provides a means for applying brakes manually without air pressure.

Helper Locomotive. A locomotive usually placed toward the rear of a train, to assist in the movement of the train over heavy grades. Diesel-electric helper locomotives can be either manned or remotely controlled from the lead unit of the train.

Hot Box Detector. A heat-sensitive device installed along main line trackage at strategic locations for measuring the relative temperatures of passing journal bearings and providing a warning to train crews if an overheated journal is discovered.

Howe Truss Bridge. A popular nineteenth-century bridge used by both public vehicles and railroads. The diagonal members slant upwards to handle compressive forces. Typically Howe truss bridges are made from wood or iron.

Janney Coupler. Resembling the knuckles of the human hand, this coupler design by the turn of the twentieth century became the standard for railway rolling stock in North America.

Journal Bearings. Plain journal bearings are blocks of metal, usually brass, shaped to fit the curved surface of the axle journal and resting directly upon it with lubrication provided by free oil contained in the journal box. Journal roller bearings are sealed assemblies of rollers and other components pressed onto axle journals and generally lubricated with grease.

Link- and Pin-Coupler. An early type of connection between freight and passenger cars that employed a link- and pin-arrangement.

Maglev Trains. These modernistic people movers operate on the principle of magnetic levitation whereby a magnetic field is created by electrified coils in the concrete guildways. By reversing polarity of the magnetized coils, thrust is created.

Mallet Locomotive. This steam locomotive, originated in the 1870s by Swiss designer Anatole Mallet, consists of a compound engine that features a jointed main frame with sets of driving wheels.

Narrow Gauge. A nonstandard gauge that usually measures 3 feet or 3 feet, 6 inches between the rail heads. *See also* Standard Gauge.

Pantograph. A device for collecting current from an overhead conductor or catenary and consisting of a jointed frame operated by springs or compressed air and having a mechanical collector at the top.

Piled Roadway. Wooden pilings and bridgelike structures used for the roadbed during the Demonstration Period.

Pilot Truck. A set of two or four unpowered wheels located under the front end of a steam locomotive. These wheels help guide the engine through curves and also support the front end.

Platform Car. A passenger-car type introduced in the latter part of the nineteenth century permitting the safe and easy movement of riders between cars. Platform construction created stronger pieces of rolling stock.

RPOs. From the 1860s to the 1970s, Railway Post Office cars (RPOs) commonly were part of passenger train consists in North America, allowing for the onboard sorting of mail.

Safety-Valve. A type of pressure relief valve, common on steam locomotives, used to protect against an accumulation of potentially dangerous excess pressure.

Slack. Unrestrained free movement between the cars of a train.

"Snake Heads." The curled-up pieces of strap rail when they become loosened or detached from the wooden rails.

Spring. A general term that refers to a large group of mechanical devices taking advantage of the elastic properties of materials to control motion or cushion loads.

Standard Gauge. The standard distance between rails (e.g., North American railroads) that measures 4 feet, 8½ inches between the inside faces of the rail heads.

Station. In common railroad usage, a station refers to the depot and all other facilities (water towns, coal chutes, and the like) at the site. Over the years, "station" and "depot" have been used interchangeably to refer to the building that houses railroad services.

Steam-Turbine Electric Locomotive. A locomotive in which steam, produced usually by coal or diesel fuel, drives electric generators that provide current to electric traction motors.

Strap Rail. Usually pieces of 1-inch-thick iron in 20- to 25-foot strips that are bolted to wooden beams to serve as rail.

Superheaters. Since the total energy contained in a given amount of superheated steam is larger than in the same amount of saturated vapor at the same boiler pressure, mechanical devices, tubes placed in the boilers of steam locomotives, are able to create this desired superheated steam.

T Rail. The standard shape of all-metal rails, introduced in the late 1830s. Early on, H and U rails were also tried.

Ties. The usually wooden component of the track structure to which rails are attached. Known also as "sills" or "sleepers."

TOFC. An acronym for "trailer-on-flatcar" intermodal service.

Traction Motor. A specially designed DC series-wound motor mounted on the trucks of modern locomotives and self-propelled cars to drive the axles.

Tractive Effort. The useable force exerted by the wheels of a locomotive at the rails for moving a train.

Trailing Truck. A set of two, four, or six normally unpowered wheels in a truck under the cab area of a steam engine. These wheels are used to provide support for the back end of the locomotive.

Tramway. A primitive guideway, commonly of wood, allowing for the movement of wagons by animal power or gravity.

Truck. The assembly of parts including wheels, axles, bearings, side frame, brake rigging, springs, and other associated components.

Unit Train. A train transporting a single commodity from one source to one destination, often coal or grain.

Vacuum Brake. A stopping system used largely outside of North America that is actuated by exhausting air from a device on each car to make use of atmospheric air to apply the brakes.

Wheel Flange. The vertical projection along the inner rim of a wheel that serves, in conjunction with the flange of the mating wheel, to keep the wheel set on the track and provides the lateral guidance system for the mounted pair.

Wooden railways. The earliest railways that consisted mostly of wooden components, including cars and track.

Wooten Firebox. Commonly found in anthracite-burning steam locomotives (like Camelbacks), this firebox is large, permitting the easy burning of hard coal.

Bibliography

Ahrons, E. L. *The British Steam Railway Locomotive, 1825–1925.* London: Locomotive Publishing Company, 1927.

Aldrich, Mark. *Safety First: Technology, Labor, and Business in the Building of Worker Safety, 1870–1939.* Baltimore, MD: Johns Hopkins University Press, 1997.

Allen, C. Frank. *Railroad Curves and Earthwork.* New York: McGraw-Hill Book Co., 1931. *The American Railway: Its Construction, Development, Management and Appliances.* Reprint, New York: Castle, 1988.

Armstrong, John H. *The Railroad: What It Is, What It Does.* Omaha, NE: Simmons-Boardman, 1982.

Bailey, Edd H. *A Life with the Union Pacific: The Autobiography of Edd Bailey.* St. Johnsbury, VT: Saltillo Press, 1989.

Barrett, Colin. *The World Encyclopedia of Locomotives.* London: Lorenz Books, 2000.

Bartky, Ian R. "The Invention of Railroad Time." *Railroad History* (Spring 1983): 13–22.

———. *Selling the True Time: Nineteenth-Century Timekeeping in America.* Stanford, CA: Stanford University Press, 2000.

Beaver, Patrick. *A History of Tunnels.* Secaucus, NJ: Citadel Press, 1972.

Beebe, Lucius. *Mixed Train Daily: A Book of Short-Line Railroads.* New York: Dutton, 1947.

Berg, Walter Gilman. *Buildings and Structures of American Railroads: A Reference Book for Railroad Managers, Superintendents, Master Mechanics, Engineers, Architects, and Students.* New York: J. Wiley & Sons, 1893.

Bezilla, Michael. *Electric Traction on the Pennsylvania Railroad, 1895–1968*. University Park: Pennsylvania State University Press, 1980.

Bianculli, Anthony J. *Trains and Technology: The American Railroad in the Nineteenth Century—Volume 1: Locomotives*. Newark: University of Delaware Press, 2001.

———. *Volume 2: Cars*. Newark: University of Delaware Press, 2002.

———. *Volume 3: Track and Structures*. Newark: University of Delaware Press, 2003.

Brignano, Mary, and Hax McCullough. *A History of Railroad Signals and the People Who Made Them*. Pittsburgh, PA: Union Switch and Signal Company, 1981.

Brill, Debra. *History of the J. G. Brill Company*. Bloomington: Indiana University Press, 2001.

Brown, John K. *The Baldwin Locomotive Works, 1831–1915*. Baltimore, MD: Johns Hopkins University Press, 1995.

Bruce, Alfred W. *The Steam Locomotive in America: Its Development in the Twentieth Century*. New York: W. W. Norton & Co., 1952.

Bryant, Keith L. Jr., ed., *Encyclopedia of American Business History and Biography: Railroads in the Age of Regulation, 1900–1980*. New York: Facts on File, 1988.

Carry, Clark E. *Railway Mail Service: Its Origins and Development*. Chicago: A. C. McClurg, 1909.

Carter, Charles F. *When Railroads Were New*. New York: Henry Holt, 1908.

Chicago, Burlington & Quincy Railroad. *Aeolus*. Chicago: Chicago, Burlington & Quincy Railroad, 1940.

Chrimes, Mike, ed. *The Civil Engineering of Canals and Railways before 1850*. Brookfield, VT: Ashgate, 1997.

Churella, Albert J. *From Steam to Diesel: Managerial Customs and Organizational Capabilities in the Twentieth-Century American Locomotive Industry*. Princeton, NJ: Princeton University Press, 1998.

Clark, Daniel K., and Zerah Colburn. *Recent Practice in the Locomotive Engine*. London: Blackie & Son, 1861.

Clark Equipment Company. *Autotram*. Battle Creek, MI: Clark Equipment Company, 1933.

Cochran, Thomas C. *Railroad Leaders, 1845–1890: The Business Mind in Action*. Cambridge, MA: Harvard University Press, 1953.

Coffin, Lorenzo S. "Safety Appliances on the Railroads." *Annals of Iowa* (January 1903): 561–582.

Condit, Carl W. *The Port of New York: A History of the Rail and Terminal System from the Beginnings to Pennsylvania Station*. Chicago: University of Chicago Press, 1980.

DeBoer, David J. *Piggyback and Containers: A History of Rail Intermodal on America's Steel Highway*. San Marino, CA: Golden West Books, 1992.

Dilts, James D. *The Great Road: The Building of the Baltimore & Ohio, The Nation's First Railroad, 1828–1853*. Stanford, CA: Stanford University Press, 1993.

Droege, John A. *Passenger Terminals and Trains*. New York: McGraw-Hill Book Co., 1916.

Drury, George. *Guide to North American Steam Locomotives*. Milwaukee, WI: Kalmbach Publishing, 1993.

Duffy, Michael C. *Electric Railways, 1880–1990*. London: Institution of Electrical Engineers, 2003.

Duke, Donald, and Edmund Keilty. *RDC: The Budd Rail Diesel Car*. San Marino, CA: Golden West Books, 1990.

Ellis, Hamilton. *Railway Carriages in the British Isles from 1830 to 1914*. London: George Allen & Unwin, 1965.

Epstein, Ralph C. *GATX: A History of the General American Transportation Corporation, 1898–1948*. New York: North River Press, 1948.

Ferguson, Eugene S., ed. *Early Engineering Reminiscences (1815–1840) of George Escol Sellers*. Washington, DC: Government Printing Office, 1965.

Fleming, Howard. *Narrow Gauge Railways in America*. Philadelphia: Howard Fleming, 1875.

Frey, Robert L., ed. *Encyclopedia of American Business History and Biography: Railroads in the Nineteenth Century*. New York: Facts on File, 1988.

Gadsden, James. *Report on the Project of a Rail Road from Charleston, S.C. to the Ohio River*. Columbia, SC: A. S. Johnston, 1836.

Gamst, Frederick C. "The Context and Significance of America's First Railroad, on Boston's Beacon Hill." *Technology & Culture* 33 (1992): 66–100.

———, ed. *Early American Railroads: Franz Anton Ritter von Gerstner's Die innern Communicationen (1841–1843)*. Stanford, CA: Stanford University Press, 1997.

Grant, H. Roger, ed. *Brownie the Boomer: The Life of Charles P. Brown, an American Railroader*. DeKalb: Northern Illinois University Press, 1991.

———. *The Corn Belt Route: A History of the Chicago Great Western Railroad Company*. DeKalb: Northern Illinois University Press, 1984.

———. *Erie Lackawanna: Death of an American Railroad, 1938–1992*. Stanford, CA: Stanford University Press, 1994.

———. *"Follow the Flag": A History of the Wabash Railroad Company*. DeKalb: Northern Illinois University Press, 2004.

———. *The North Western: A History of the Chicago & North Western Railway System*. DeKalb: Northern Illinois University Press, 1996.

———. "Piggyback Pioneer: The Chicago Great Western Railroad." *Trains* (January 1986): 31–34.

———, ed. *We Took the Train*. DeKalb: Northern Illinois University Press, 1990.

———, and Charles W. Bohi. *The Country Railroad Station in America*. Boulder, CO: Pruett Publishing, 1978.

———, Don L. Hofsommer, and Osmund Overby. *St. Louis Union Station: A Place for People, A Place for Trains*. St. Louis: St. Louis Mercantile Library, 1994.

Haupt, Hermon. *General Theory of Bridge Construction*. New York: D. Appleton & Co., 1851.

Hay, Warren H. "The Beginnings of Telephone Dispatching." *Railroad History* (Spring 1974): 55–60.

Hilton, George W. *American Narrow Gauge Railroads*. Stanford, CA: Stanford University Press, 1990.

———, and John F. Due. *The Electric Interurban Railways in America*. Stanford, CA: Stanford University Press, 1960.

Hirsimaki, Eric F. *Lima: The History*. Edmunds, WA: Hundman Publishing, 1986.

Hofsommer, Don L., ed. *Railroads in Oklahoma*. Oklahoma City: Oklahoma Historical Society, 1977.

———. *The Southern Pacific, 1901–1985*. College Station: Texas A&M University Press, 1986.

Hollingsworth, Brian, ed. *The Illustrated Encyclopedia of North American Locomotives: A Historical Directory of America's Greatest Locomotives from 1830 to the Present Day*. New York: Crescent Books, 1984.

Huddleston, Eugene L. *Uncle Sam's Locomotives: The USRA and the Nation's Railroads*. Bloomington: Indiana University Press, 2002.

Hunter, Robert F., and Edwin L. Dooley Jr. *Claudius Crozet: French Engineer in America, 1790–1864*. Charlottesville: University Press of Virginia, 1989.

Jackson, William. *Lecture on Rail Roads*. Boston: Henry Bowen, 1829.

Jensen, Oliver. *The American Heritage History of Railroads in America*. New York: Bonanza Books, 1981.

Johnson, Ralph P. *The Steam Locomotive*. New York: Simmons-Boardman, 1942.

Kaminski, Edward S. *American Car & Foundry Company, 1899–1999*. Wilton, CA: Signature Press, 1999.

Keilty, Edmund. *Doodlebug Country*. Glendale, CA: Interurbans, 1982.

———. *Interurbans without Wires*. Glendale, CA: Interurbans, 1979.

Kidner, R. W. *The Development of the Railcar*. South Godstone, England: Oakwood Press, 1958.

Kiefer, Paul W. *A Practical Evaluation of Railroad Motive Power*. New York: Simmons-Boardman, 1949.

Klein, Maury. "Replacement Technology: The Diesel as a Case Study." *Railroad History* (Spring 1990): 109–120.

———. *Union Pacific: The Birth of a Railroad, 1862–1893*. Garden City, NY: Doubleday & Company, 1987.

———. *Union Pacific: The Rebirth, 1894–1969*. New York: Doubleday, 1989.

Kube, Kathi. "Get Ready for No-Man Trains." *Trains* (September 2003): 42–51.

Lamb, J. Parker. *Perfecting the American Steam Locomotive*. Bloomington: Indiana University Press, 2003.

Lamb, Martha J. "Glimpses of the Railroad in History." *Magazine of American History* (June 1891): 429–447.

Larkin, F. Daniel. *John B. Jervis: An American Engineering Pioneer*. Ames: Iowa State University Press, 1990.

LeMassena, Robert A. *Articulated Steam Locomotives of North America*. Silverton, CO: Sundance Publications, 1979.

Lewis, M.J.T. *Early Wooden Railways*. London: Routledge and Kegan Paul, 1970.

MacGregor, Bruce. *The Birth of California Narrow Gauge: A Regional Study of the Technology of Thomas and Martin Carter.* Stanford, CA: Stanford University Press, 2003.

Marre, Louis A. *Diesel Locomotives: The First 50 Years.* Waukesha, WI: Kalmbach Publishing Co., 1995.

Martin, Albro. *Enterprise Denied: Origins of the Decline of American Railroads, 1897–1917.* New York: Columbia University Press, 1971.

Meeks, Carroll L. V. *The Railroad Station: An Architectural History.* New Haven, CT: Yale University Press, 1956.

Melnikov, Colonel. *Technical Description of the Railways in the United States.* N.p., n.d.

Meunier, Jacob. *On the Fast Track: French Railway Modernization and the Origins of the TGV, 1944–1983.* Westport, CT: Praeger Publishers, 2000.

Middleton, William D. *Landmarks on the Iron Road: Two Centuries of North American Railroad Engineering.* Bloomington: Indiana University Press, 1999.

———. *When the Steam Railroads Electrified.* Milwaukee: Kalmbach Publishing Co., 1974.

Misa, Thomas J. *Nation of Steel: The Making of Modern America, 1865–1925.* Baltimore, MD: Johns Hopkins University Press, 1995.

Moody, Linwood W. *The Maine Two-Footers: The Story of the Two-Foot Gauge Railroads in Maine.* Berkeley, CA: Howell-North, 1959.

Morgan, David P. "Faith in Steam." *Trains* (November 1954): 18–30.

———. "N&W's Tomorrow." *Trains* (November 1954): 29–30.

———. "The Semmering Story." *Trains* (October 1961): 43–49.

———. *Steam's Finest Hour.* Milwaukee, WI: Kalmbach Publishing Co., 1959.

New-York & Erie Railroad Company. *Second Report of the Directors of the New-York & Erie Railroad Company to the Stockholders.* New York: Egbert Hedge, 1841.

Nock, O. S. *The Dawn of World Railways, 1800–1850.* New York: Macmillan, 1972.

Overton, Richard C. *Burlington Route: A History of the Burlington Lines.* New York: Knopf, 1965.

"Pace that Kills." *World's Work* 13 (March 1907): 8, 595–596.

Pennsylvania Railroad. *Rules of Operation.* Philadelphia: Pennsylvania Railroad, 1849.

Postman, Neil. *Technopolgy: The Surrender of Culture to Technology.* New York: Knopf, 1992.

Reck, Franklin M. *The Dilworth Story: The Biography of Richard Dilworth Pioneer Developer of the Diesel Locomotive.* New York: McGraw-Hill, 1954.

Richards, Jeffrey, and John M. MacKenzie. *The Railway Station: A Social History.* New York: Oxford University Press, 1986.

Ross, David, ed. *The Encyclopedia of Trains and Locomotives.* San Diego, CA: Thunder Bay Press, 2003.

Solomon, Brian. *Modern Locomotives: High-Horsepower Diesels, 1996–2000.* St. Paul, MN: MBI Publication Co., 2002.

Stover, John F. *American Railroads.* 2nd. ed. Chicago: University of Chicago Press, 1997.

———. *History of the Baltimore and Ohio Railroad.* West Lafayette, IN: Purdue University Press, 1987.

Taylor, George Rogers. *The Transportation Revolution, 1815–1860.* New York: Holt, Rinehart and Winston, 1951.

———, and Irene D. Neu. *The American Railroad Network, 1861–1890.* Cambridge, MA: Harvard University Press, 1956.

"10 Diesel Locomotives that Most Changed Railroad." *Trains* (October 2002): 36–47.

Tichi, Cecilia. *Shifting Gears: Technology, Literature, and Culture in Modernist America.* Chapel Hill: University of North Carolina Press, 1987.

Trostel, Scott D. *The Barney & Smith Car Company: Car Builders.* Fletcher, OH: Cam-Tech Publishing, 1993.

The Truth about the Railroads. St. Paul, MN: 1913.

Usselman, Steven W. *Regulating Railroad Innovation: Business, Technology, and Politics in America, 1840–1920.* New York: Cambridge University Press, 2002.

Vance, James E. Jr. *The North American Railroad.* Baltimore, MD: Johns Hopkins University Press, 1995.

Vose, George L. "Notes Relating to the Early History of Transportation in Massachusetts." *Journal of the Association of Engineering Societies* 4 (1884): 67–68.

Waite, Thornton. "Dr. Borst's X-12: The Atomic Locomotive." *Railroad History* (Autumn 1996): 37–55.

Ward, James A. *Railroads and the Character of America, 1820–1887.* Knoxville: University of Tennessee Press, 1986.

———. *That Man Haupt: A Biography of Herman Haupt.* Baton Rouge: Louisiana State University Press, 1973.

Warden, William E. Jr. "Claudius Crozet: Napoleon's Capitain versus the Blue Ridge." *Railroad History* (Autumn 1973): 44–55.

Westing, Frederick. *The Locomotives That Baldwin Built.* New York: Bonanza Books, 1966.

Westinghouse Electric Company. *Westinghouse Electric Railway Transportation.* Chicago: Central Electric Railfans' Association, Bulletin 118, 1978.

Westwood, J. N. *Locomotive Designers in the Age of Steam.* Rutherford, NJ: Fairleigh Dickinson University, 1978.

White, John H. Jr. *American Locomotives: An Engineering History, 1830–1880.* Baltimore, MD: Johns Hopkins University Press, 1997.

———. *The American Railroad Freight Car: From the Wood-Car Era to the Coming of Steel.* Baltimore, MD: Johns Hopkins University Press, 1993.

———. *The American Railroad Passenger Car.* Baltimore, MD: Johns Hopkins University Press, 1978.

———. *The Great Yellow Fleet: A History of American Railroad Refrigerator Cars.* San Marino, CA: Golden West Books, 1986.

———. *The John Bull.* Washington, DC: Smithsonian Institution Press, 1981.

———. *On the Right Track: Some Historical Cincinnati Railroads.* Cincinnati, OH: Cincinnati Railroad Club, 2003.

———. *A Short History of American Locomotive Builders in the Steam Era.* Washington, DC: Bass, 1982.

Withuhn, William, ed. *Rails across America: A History of Railroads in North America.* New York: Smithmark Publishers, 1997.

Wright, Benjamin. *Considerations on the Subject of the New-York and Erie Railroad.* N.p., 1833.

Young, Andrew D., and Eugene F. Provenzo Jr. *The History of the St. Louis Car Company.* Berkeley, CA: Howell-North Books, 1978.

Index

About the Author

H. ROGER GRANT is professor of history at Clemson University. He is a specialist in American transportation history. Some of his recent books include histories of the Erie Lackawanna, Chicago & North Western and Wabash railroads and he is completing a book-length study of the Georgia & Florida Railroad.